FOSTERING FUNDAMENTALISM

US Foreign Policy and Conflict in the Islamic World

Series Editors:
Tom Lansford
The University of Southern Mississippi – Gulf Coast, USA
Jack Kalpakian
Al-Akhawayn University, Morocco

The proliferation of an anti-US ideology among radicalized Islamic groups has emerged as one of the most significant security concerns for the United States and contemporary global relations in the wake of the end of the Cold War. The terrorist attacks of September 11, 2001 demonstrated the danger posed by Islamic extremists to US domestic and foreign interests. Through a wealth of case studies this new series examines the role that US foreign policy has played in exacerbating or ameliorating hostilities among and within Muslim nations as a means of exploring the rise in tension between some Islamic groups and the West. The series provides an interdisciplinary framework of analysis which, transcending traditional, narrow modes of inquiry, permits a comprehensive examination of US foreign policy in the context of the Islamic world.

Other titles in the series:

US-Pakistan Relationship
Soviet Invasion of Afghanistan
A.Z. Hilali
0 7546 4220 8

Strategic Preemption
US Foreign Policy and the Second Iraq War
Robert J. Pauly, Jr. and Tom Lansford
0 7546 4357 3

Eurasia in Balance
The US and the Regional Power Shift
Ariel Cohen
0 7546 4449 9

Fostering Fundamentalism
Terrorism, Democracy and American Engagement in Central Asia

MATTHEW CROSSTON
Clemson University, USA

ASHGATE

Published by
Ashgate Publishing Limited
Gower House
Croft Road
Aldershot
Hants GU11 3HR
England

Ashgate Publishing Company
Suite 420
101 Cherry Street
Burlington, VT 05401-4405
USA

Ashgate website: http://www.ashgate.com

British Library Cataloguing in Publication Data
Crosston, Matthew
Fostering fundamentalism : terrorism, democracy and American engagement in Central Asia. - (US foreign policy and conflict in the Islamic world)
1.War on Terrorism, 2001- 2.United States - Foreign relations - Asia, Central 3.Asia, Central - Foreign relations - United States 4.United States - Foreign relations - 2001- 5.Asia, Central - Foreign relations - 1991- 6.Asia, Central - Politics and government - 1991 - 7.Asia, Central - Social conditions - 1991 -
I.Title
327.7'3'058'0090511

Library of Congress Cataloging-in-Publication Data
Crosston, Matthew.
Fostering fundamentalism : terrorism, democracy, and American engagement in Central Asia / by Matthew Crosston.
p. cm. -- (US foreign policy and conflict in the Islamic world)
Includes bibliographical references and index.
ISBN-13: 978-0-7546-4632-7
ISBN-10: 0-7546-4632-7
1. United States--Foreign relations--Asia, Central. 2. Asia, Central --Foreign relations--United States. 3. Asia, Central--Politics and government--21st century. 4. Democracy. 5. National security --United States. 6. Terrorism--United States--Prevention. I. Title.
JZ1480.A55C76 2006
327.73058--dc22

2006018446

ISBN-13: 978 0 7546 4632 7
ISBN-10: 0 7546 4632 7

Printed and bound in Great Britain by MPG Books Ltd. Bodmin, Cornwall.

Contents

List of Maps, Tables and Figures

It is forbidden to kill.
Therefore all murderers are punished.
Unless they kill in large numbers and to the sound of trumpets.

Voltaire

For all those who wish to silence the trumpets.

Chapter 1

Introduction: The Global War on Terror and Emerging Threats

I want to speak to the people of Central Asia: The United States believes that liberty and dignity and justice are within reach of everyone in this region. And we are fully committed to partnership in helping you to realize this vision. We seek peace and security. We seek economic development and prosperity. We seek democratic values and human rights that unite all free nations in trust and in respect.[1]

Condoleeza Rice,
US Secretary of State

How do we guarantee 9/11 was and will be a unique incident, never to be repeated? The United States has adopted a dual foreign policy prioritization in the twenty-first century to answer this question. It must end terrorist groups wherever they are to be found. It must seed democratic regimes so that terror groups will not find fertile soil in which to reemerge. It is sincerely believed within the corridors of power that these dual objectives can be achieved simultaneously. It is stridently ignored when suggested these two objectives might actually collide with one another. This book is ultimately about that collision, arguing for the essential prioritization of long-term security strategies that focus on legitimate democratic development, *even if it may appear to compromise the short-term fight against terror*. This analysis takes to task a disturbing trend that seems to verbally endorse long-term strategies but then consistently undermines them in real-time with short-term policies of convenience. It will reveal to the reader why the sincere effort to fight the Global War against Terror today often results in the US compromising its principles on democratic development in such a way as to give cause for future terror to grow. Our dual foreign policy priorities *do* collide. The aftermath of that collision makes us less safe because our 'engagement' too often violates the essence of what it is to be American: democratic, stable, open and just. Most importantly of all, this analysis goes beyond the standard moral vs. national interest debates. *Fostering Fundamentalism* argues that the Global War against Terror offers the US a unique opportunity: a situation where doing what is moral and just is actually the best way to serve American security.

This runs contrary to traditional takes in international politics. Protecting physical security has always been considered a more immediate national interest, while democratization is termed a benefit to the United States only indirectly, producing results much further down the political road. More contrary still is my supposition that the present fight against terror (think Iraq and Afghanistan) is not actually the greatest danger fueling radical fires to burn against the United States. What is the ultimate danger in the Global War against Terror emerges from the continued compromises made on democratic development with those governments who move further away from freedom while we publicly praise ourselves, declaring our engagement to have moved them ever closer to liberty. This subtle hypocrisy radicalizes local opposition not only against the indigenous governments but against all foreign powers who duplicitously support such regimes. This inevitably leads to an emerging hatred against the United States best expressed through the ruthless and cruel methods of terror.

Most scholars who have argued similar veins in the past are open to the criticism of unrealistic idealism and an unwillingness to accept how real statecraft and foreign policy are conducted across the globe. This counterargument always cites the overarching necessity to secure power and reduce threats, making short-term compromises, though perhaps hypocritical, ultimately inevitable. This investigation is different because it speaks sincerely in the language of the hawks: we increase our threats and make our power less secure through these compromises. This is not statecraft or realpolitik. It is short-sighted and deadly.

In short, this is a treatise arguing against dangerous foreign policy stupidity and shows the reader just how that danger emerges. This is about stopping contradictions in our foreign policy that will only compromise American lives down the road. It is hoped that the reader will therefore see the pragmatism running though this work and will not waste time with accusations of idealism and naiveté. Security is more important to the final conclusions here than global happiness and understanding. Indeed, the resolution to this problem I argue is self-evident though not easy: the United States should actually practice what it preaches at the State Department because the fact of the matter is our sermons are more right and more strategically advantageous than our acts. It is time to achieve some synergy between policy and practice.

A Brief Overview of the Literary and Theoretical Landscape

This is not a purely intellectual discussion of liberalism vs. realism, multilateralism vs. unilateralism, diplomacy vs. intervention or even the nurture vs. nature debate within democracy circles. These theoretical debates, all vitally important, are not the foci of this investigation though they clearly inform the directions taken within. It is my hope that *Fostering Fundamentalism (FF)*, though pragmatic in its overarching analysis of foreign policy, will earn a spot amongst the fine works that address such debates. We all hang on the branches of one academic family tree.

Perhaps the most obvious comparison with a recent work that takes on similar themes is Zakaria's *The Future of Freedom*. An important and impressive work, *The Future of Freedom* calls for a restoration of the balance between liberty and democracy and shows how liberal democracy has to be made effective and relevant for our time.[2] With this I can do nothing except wholly agree. In fact, *Fostering*

Fundamentalism does its best to show how American foreign policy is failing to make liberal democracy more effective and relevant. Indeed, it shows how the US even undermines its own effort to achieve such goals. Where I perhaps respectfully step away from Zakaria's treatise is in his discussion dealing with liberal constitutionalism. Paraphrasing crudely, Zakaria cites how those countries that had the benefit of time and gradual change were able to develop the institutions of democracy more fervently because their societies were already at least partially open to the attitude that infuses the institutions, that is, liberal constitutionalism.

I disagree with Zakaria's premise in so much as I do not accept it as an irrefutable given for all future democratization. I believe the United States can with its power and its own liberal constitutionalism practice a foreign policy that may help engender the development of such constitutionalism in other places. I am not sure Zakaria believes that is possible. I believe it is and it informs the optimism underneath the warning embedded within *Fostering Fundamentalism*.

Zakaria's book I believe to be more balanced and objective than many of the other works that inform this highly topical debate about democracy and terror. Amy Chua's *World on Fire*, for example, fascinatingly argues that America today enjoys a wealth and economic power wildly disproportionate to its numbers. This accounts for the visceral hatred of Americans expressed in recent acts of terrorism. Chua warns that, far from making the world a better and safer place, democracy and capitalism – at least in the raw, unrestrained form in which they are currently being exported – are intensifying ethnic resentment and global violence with potentially catastrophic results.[3]

Interestingly, I think Chua documents empirical cases that are the consequence of the hypocrisy *FF* seeks to reveal and argue against. Perhaps where we differ most strongly is the sense that Chua may be a proponent of the theory that such hypocrisy is by design, that democracy and market transition in and of itself can only produce such horrendous inequality, suffering and strife. I do not belong to what is ultimately a quasi-Marxist frame of reference. *Fostering Fundamentalism* does have one theoretical leap of faith that works like Chua's would stridently disagree with: since I examine a flawed application of democratic support, it is theoretically available for me to believe in positive results where democratic engagement is done properly. This analysis aims to show how this is true. Fine authors like Chua would, I believe, simply agree to disagree with my optimism.

Arguably the most well-known intellectual cut from this cloth would be Noam Chomsky. His work *Hegemony or Survival* dissects America's quest for 'global supremacy', tracking the U.S. government's aggressive pursuit of policies intended to achieve 'full spectrum dominance' at any cost. He lays out vividly how the various strands of policy cohere in a drive for hegemony that ultimately threatens our survival. In our era, he argues, empire is a recipe for an earthly wasteland.[4]

As with Chua, it isn't so much that *Fostering Fundamentalism* seeks to discredit or dismiss the biting logic of Chomsky. It simply refuses to buy into the conspiracy theory that the US Department of State purposefully seeks to undermine other countries with a hypocritical democratization project. *FF* backs away from these positions because it shows how such hypocrisy actually *undermines and subverts*

American security and power, rather than enhancing and intensifying it. Simple realist logic argues against states taking actions that ultimately compromise its position on the global stage. The deadly mistakes Chomsky discusses are real. They are emblematic, however, of a failure to remain true to our actual democratic mission rather than being the perverse fruition of a realized foreign policy goal.

Tipping the intellectual scales back in the other direction are works like *Statebuilding*, by Francis Fukuyama. He argues the well-governed polity was always a necessary precondition as 'weak or failed states are the source of many of the world's most serious problems'. He traces what we know and more often don't know about how to transfer functioning public institutions to developing countries in ways that will leave something of permanent benefit to the citizens of the countries concerned. He ends the book with a discussion of the consequences of weak states for international order and the grounds on which the international community may legitimately intervene to prop them up.[5]

Perhaps somewhat surprisingly, *FF* is not completely in line with scholars like Fukuyama. While *FF* speaks the language of the hawks it is not slavishly supportive of hawkish motives: the ultimate increase in long-term security to America with the real enlargement of the cradle of democracy is not because it supports our dominance but because it supports the stability and prosperity for others and gives them the institutions and tools necessary to air and resolve grievances. Conservative scholars like Fukuyama can overemphasize the militaristic applications to 'legitimate intervention' and can too often understand short-term rationalizations that have the US in bed with dictators and tyrants. *FF* analyzes the long-term security dangers that are inevitable to such actions. As such, works like *Statebuilding*, while important and necessary, are not automatically brothers-in-intellectual-arms with *FF*.

Finally, Natan Sharansky's *The Case for Democracy* rounds out the academic field where *FF* firmly sits. Sharansky takes non-democratic societies and puts them under a microscope to reveal the mechanics of tyranny that sustain them. In exposing the inner workings of a 'fear society', Sharansky explains why democracy is not beyond any nation's reach, why it is essential for our security and why there is much that can be done to promote it around the world.[6]

There is much within *FF* that Sharansky would agree with. We both agree with the transformative power of real democracy and lament its uneven and hypocritical application as well as our ability to justify policy initiatives that act against our philosophical founding. Sharansky, however, seems more willing to accept this aspect of foreign policy as unchangeable and focuses more on the need to push aggressively for the elimination of 'fear societies', replacing them with free ones. I believe Sharansky would not be pleased with the length of time necessary to produce change as argued within *FF* because it does not argue as such for military intervention. It is not so much that I believe military intervention is unacceptable. It is simply that *FF* shows why any military intervention, as presently conducted, will be doomed to long-term failure exactly because we have not yet rectified the hypocrisy within our foreign policy and diplomatic engagement. As long as it remains this way then Islamic radicals will always have a counter against our supposed 'good will' that we cannot legitimately refute. We do fight some tyrannies. We also support others. How do we argue against such a brazen rejection of who we are? We cannot, which is the entire point of *FF*.

These works are by no means the only representatives in the ever-enlarging field to examine democracy and terror together. What they do represent, however, is an appropriate sampling of the full spectrum, with authors like Chua and Chomsky on the left, Fukuyama and Sharansky on the right, and Zakaria and myself (if I may be allowed to be so bold as to include myself in such an esteemed group) in the middle. Taken together, these six works will give a comprehensive analysis of this complex foreign policy problem and also give readers a feel for the intellectual latitude that exists amongst the scholars themselves.

With Friends Like These

This book takes the reader to the isolated Ferghana Valley in Central Asia. It gives readers an intimate look into three states very few have heard of and reveals political processes few experts have realized. It shows how our supposed support of democratization dies in the Valley, replaced by the diplomatic necessity of compromise. Unfortunately, that compromise may ultimately kill us all.

The present work analyzes US foreign policy in three of the five Central Asian states: Uzbekistan, Tajikistan, and Kyrgyzstan. The Ferghana Valley is a relatively fertile zone that stretches awkwardly across parts of northern Tajikistan, south-western Kyrgyzstan and eastern Uzbekistan.

Map 1.1 The Ferghana Valley

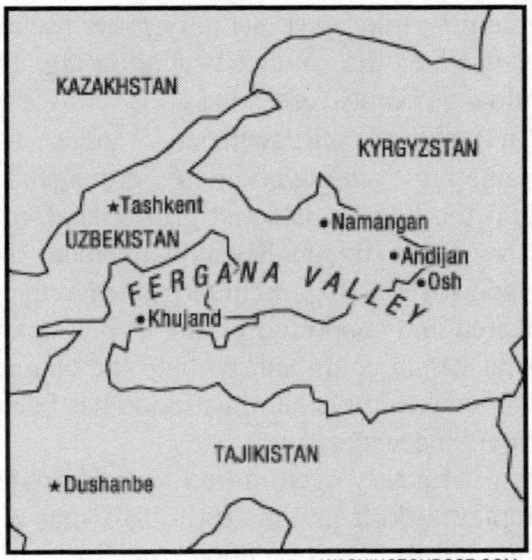

WASHINGTONPOST.COM

From a purely cultural perspective, Uzbekistan has always considered the Ferghana Valley an intrinsically Uzbek area. To a large extent this is true. However, when speaking in terms of political geography, the fact of the matter is that the valley is split between the three states. When three states intersect over one area and are given the charge of administering and governing it cooperatively, the old American football adage comes into play: if you have three quarterbacks playing on your team, you have *no* quarterbacks on your team. As a result, the effective management of the Ferghana Valley has remained an amorphous and ambiguous proposition since the dissolution of the Soviet Union. In fact, as a closer look is offered at these three 'allies' in the war against terror it becomes necessary to wonder: with friends like these who needs enemies?

Uzbekistan

In May 2002 the Uzbek government officially abolished its long-held practice of state censorship. In a state that has had a vice-like grip over all sources of media since it became independent in 1991, such a declaration should have been greeted with universal applause. The move was undermined, however, by the warning that accompanied the announcement and was given to all the editors in the country about their 'responsibility' for content.[7]

The chairman of the State Press Committee, Rustam Shagulyamov, explained that all editors would now be accountable for everything they published. While it was not revealed what consequences would emerge were an editor to publish something the government deemed 'improper', members of the Uzbek mass media did not miss that the State Secrets Inspectorate, which had for years been charged with exclusively censoring newspapers and defining editorial policies, did not suffer a major change in mission despite the 'abolishment' of state censorship. In other words, a day trumpeted in the West as a signal of true progress toward freedom of expression was seen in Uzbekistan as a potentially ominous forewarning: the government may have actually *freed* its hands to act with more impunity by relieving itself of the official watchdog tag. It no longer had to be preventative. It could now be purely reactionary.

In September 2002 a Tashkent court sentenced Yuldash Rasulov to seven years in prison for disseminating 'antigovernment' propaganda and for allegedly recruiting members into the banned Islamist group, Hizb-ut-Tahrir.[8] In actuality, Rasulov was a member of the Human Rights Movement of Uzbekistan and had been working to persuade several Hizb members to leave the Islamist organization. The prosecution, protected and supported by the Uzbek government, produced no evidence during the trial which could substantiate the charges. The subordination of the judicial branch to the executive administration has been a widely-known fact across Uzbekistan for over a decade.

The Uzbek parliament, arguably even more of a rubber-stamp for the executive than the judiciary, adopted a draft law in April 2003 that effectively rendered all former presidents immune from prosecution for life. Under the law, former presidents became automatic members of the Senate. This senatorial appointment had no term-limits and no requirement to actually *run* for office. The post was symbolically bestowed upon the president.[9] The logic behind this initiative was supposedly to be able to tap into the former presidents' knowledge of state affairs and promote Uzbekistan's development, even after the president had left office. Of course, it seems fairly clear that the draft law was meant more as a gift to prevent future prosecution against what was an administration rife with corruption.

Human rights groups have consistently estimated the number of political and religious prisoners stuck in Uzbek jails without charges or trial dates to be near 6,500. The number could very likely be higher. A UN rapporteur revealed that torture by security forces appeared to be systematic.[10] The protests over human rights raised by the United States with the Uzbek government have consistently remained along lower level diplomatic channels and have never compromised the larger priority of Uzbek cooperation in the US war against terror.

The government in early 2004 began an initiative to tighten control over international human rights and democracy organizations, claiming these groups often hid secret revolutionary agendas and aimed to create a revolution within Uzbekistan.[11] It created a new requirement demanding that all international groups must register with the Justice Ministry. Previously, international groups were under the exclusive jurisdiction of the Foreign Ministry. The new process meant to give the Justice Ministry (where all of the Uzbek security services are housed) greater oversight capacity. In effect, it granted any agency in the government the right to veto a foreign group's registration and subversively monitor its operations.

Tajikistan

The main Tajik opposition party, the Islamic Renaissance Party of Tajikistan (IRPT), accused the government of systematically persecuting party leaders and randomly arresting its supporters under the guise of preventing the spread of radical Islamism. This was already widespread *before* 9/11.[12] The IRPT declared that such systemic abuse would only risk reprisals that would increasingly grow more desperate and violent, ultimately destabilizing what was an already shaky political situation.

President Rakhmonov strengthened his long-term grip on power in the summer of 2003 when he pushed forward amendments to be voted on by popular referendum. The amendments allowed him to stand for two more seven-year terms after his current mandate expired in 2006.[13] The provisional results revealed a Soviet-like 93 percent of voters approving the amendments with nearly 97 percent of all eligible voters supposedly taking part. Though taking place in the 21st century, Rakhmonov's referendum was a rubber-stamp inside job reminiscent of the patently fake 'show elections' common during Soviet times. Asliddin Sokhibazarov and Shokir Khakimov, deputy heads of the Social Democratic Party, declared the whole initiative a farce, with documented evidence of individuals voting for every member of their family, thus casting numerous votes at a single time. Fake or not fake, pushing the amendments through popular referendum gave Rakhmonov an air of democratic legitimacy.

In spring 2004 another Tajik opposition party, Taraqqiyot, was prevented from registering for upcoming elections by the Justice Ministry.[14] The official reason for the rejection was falsified information regarding party membership. *Eleven* of the 1,173 people listed as party supporters of Taraqqiyot denied involvement in the party. Taraqqiyot chairman Sulton Quvvatov claimed those eleven members were illegally detained and ultimately tortured to force them to renounce their political affiliation. Taraqqiyot has remained within the legal system for now, having appealed the decision to the Tajik Supreme Court and threatening to take their case to the International Court of Justice in The Hague. But given such appeals are likely to be unsuccessful, it is necessary to wonder how long such obvious repression and violation of democratic rights will continue until they move underground and radicalize their positions?

This seemed to be the essence of the message of US Senator John McCain, who expressed 'outrage' over parliamentary elections in both Kyrgyzstan and Tajikistan, decrying how 'election day was marred by vote buying, de-registration of candidates, media interference, and government hyperbolic warnings about the dangers of civil war'.[15] McCain was amazed at the level of 'information control' that was rampant within both states, not seen since the worst of Soviet days. McCain chastised Kyrgyzstan and claimed the abuses 'were even worse' in Tajikistan. 'These two countries can pretend no longer', railed McCain, 'the governments in these countries present a false choice – chaotic democracy or authoritarian stability'.

McCain's consistent concerns echo throughout the present work. In a way it is unfortunate that he is the lone voice within American government who seems to care. While McCain has built a well-deserved and admirable reputation he remains a maverick known for diplomatically insensitive outbursts. This makes him a favorite outside the beltway but severely compromises his political impact on actual executive policy. Unfortunately, having his lone voice has not been a guarantee to bring changes within the American corridors of power.

Kyrgyzstan

By the time of publication, Kyrgyzstan will only be little more than a year removed from its so-called 'democratic revolution' in which the entrenched former autocrat, Askar Akaev, was forced to flee into exile. This was supposedly the latest in a line of governmental turnovers that began in the Caucasus two years ago and migrated east. It is considered by the west as further confirmation of the inexorable thirst all people have for democracy and a testimony to US foreign policy correctness. The formal chapter that deals with Kyrgyzstan reveals why this characterization is far from the truth. Meanwhile, problems have already begun to emerge in Kyrgyzstan, despite the fact that the official afterglow of the 'democratic revolution' has barely had time to die down.

Only three months after Akaev's removal police in the capital, Bishkek, had to use tear gas and batons to disperse protestors marching toward government square. The demonstrators were protesting in support of a candidate who had been barred by the new government from participating in July's presidential poll.[16] Kurmanbek Bakiev became acting president (and later would win the presidential election) after a wave of protests following controversial parliamentary elections. Bakiev emphasized that he was not against using force to quell protests, conveniently labeling them all as Akaev supporters seeking to disrupt his revolution.

Police and security services followed up the dispersal of the protest with strategic arrests against Mukar Cholponbaev, a former speaker of Kyrgyzstan's parliament, and another would-be presidential candidate Baryktabasov.[17] Omurbek Subanaliev, the Bishkek police chief, accused the two of financing and organizing the protest. The charges appeared to be politically motivated so as to curtail the only real challengers to Bakiev in the upcoming presidential elections.

In October 2005 the chairman of the Kyrgyz parliamentary Committee on State System and Law, Kambaraly Kongantiyev, began maneuvers to legally create a year-long moratorium on protests.[18] 'The situation is unstable', declared Kongantiyev in parliament, 'the moratorium is inevitable. It does mean a *deviation from democracy* or a violation of civil rights and freedoms, [but] it is an objective necessity and the only way to stabilize the situation in the country'. The official legal process for such a maneuver is for parliament to draft the resolution and then submit it to the president for approval, which is quite fortuitous considering most of the protests threatening Kyrgyz society revolve around new President Bakiev.

Amidst this environment of 'Kyrgyz democracy' stood the OSCE special envoy Alojz Peterle, who met with First Deputy Prime Minister Kulov in summer 2005. Peterle commented, 'I am more than pleased that the [Kyrgyz presidential] elections were done in that way as they were done'.[19] In addition, the OSCE envoy felt that the new Kyrgyz government had been preparing themselves to do 'what was necessary' to benefit the country. Considering such statements fly in the face of empirical fact, it is hardly surprising why conspiracy theories emerge amongst radical opposition groups, blaming the international community in general and the United States in particular as being complicit in all domestic violations of civil liberties and human rights.

Emancipators or Enablers: The Consequences of US Engagement

To a large extent Central Asia was an afterthought compared to the Russian Federation with the dissolution of the Soviet Union. Russia's size, historical relevance and nuclear arsenal gave it a guarantee of remaining globally significant, regardless of how the post-Soviet democratic transition went. As long as that transition remained a major US priority (ensuring that the 'peace dividend' from the end of the Cold War remained long-term was the type of phrase often used throughout the early to mid 1990s), then Central Asian states would also indirectly benefit from some of the diplomatic and foreign aid overflow to the region. As Russia's democratic transition began to shift more towards quasi-authoritarian centralization and oligarchic monopolies, the entire post-Soviet space suffered the consequential American political backlash. Not only did aid, assistance and attention dwindle towards Russia, it also lessened across Central Asia. Of course, this trend would shift dramatically one clear sunny day in New York City in September 2001.

Clearly 9/11 is the crescendo moment in changing President Bush's initial take on foreign policy. It is easy to forget that it was Bush in 2000 who openly criticized Gore for the likelihood of his administration being *too active* abroad and forcing American commitments across too many areas. 9/11 supposedly turned Bush from being a quasi-isolationist president into one firmly standing on a foundation of preemption: hit threats early and overseas so that they have limited opportunities to hit you later and at home. When the Bush administration finally defined this new vision of justified military intervention, based on the idea that the US would hit terror centers before they would have a chance to conduct operations on US soil, it was immediately obvious that Central Asia in general and the three states that share the Ferghana Valley in specific would play a crucial strategic role. It was no small matter that directly south of Tajikistan were the northern territories of Afghanistan, home to the largest rebel faction willing to fight against the Taliban and for a long time the archenemy of Osama bin Laden.

Elizabeth Jones, head of the State Department's Bureau of European and Eurasian Affairs, testified to this fact as early as December 2001 when she described the US vision for the region as 'serious and for the long-term'. She hoped to see a stable, peaceful and prosperous region where individual countries 'markedly accelerated their economic reforms and democratic credentials, respected human rights, and developed vibrant civil societies'.[20] This vision of Central Asia, she said, saw an increasing integration into the global economy 'via an east-west corridor of cooperation stretching from China and Afghanistan across the Caucasus to the Mediterranean'. Attempting to deflect the timing of this intensified interest away from purely strategic security concerns, Jones reiterated to Congress that, 'we are ready to explore new areas of assistance…but only in exchange for demonstrated, concrete steps toward reform'.

Two years later Jones and the State Department had not changed their tune, emphasizing the continued commitment and focus of the US on not simply using the Central Asian states for their proximity to Afghanistan and other potential Middle East launching points, but on encouraging them to engage in real democratic reform:

> When we talk with leaders of Central Asian countries, we always remind them of the need to do a better job of living up to their own promises as well as international commitments to democratic pluralism and economic openness. We emphasize the centrality of such reforms to long-term stability. Our vision is simple: namely, that these countries remain independent and become democratic, stable and prosperous partners of the United States who respect human rights, are increasingly integrated into the global economy, and avoid the poverty, isolation, and intolerance that breed terrorism and fundamentalism.[21]

The preceding anecdotal evidence from Uzbekistan, Tajikistan and Kyrgyzstan raises concern as to the sincerity of the State Department's vision for Central Asia.

If that vision, committed as it was to supposedly seeing a democratically plural and economically open Central Asia, was false, it becomes crucially necessary in today's world of global terror to understand the consequences. Anecdotes, while interesting to read, do not provide analytical evidence however. Only a more intensive and rigorous study of US foreign policy, along with an analysis of the inner workings of these three states and the emergence of radical Islamist groups in and around the Ferghana Valley, can provide that evidence. This is the objective of *Fostering Fundamentalism.*

Chapter Two is an in-depth look at US foreign policy and analyzes how the United States has actually engaged this region of Central Asia since 9/11 and whether that engagement actually does sow the seeds for new democracy. Chapter Three is an overview for the reader to get acquainted with both Central Asia in general and the Ferghana Valley in particular, letting them understand why this area is supremely positioned to be a future nexus of terror. Chapters Four, Five, and Six are individual chapters on Kyrgyzstan, Tajikistan, and Uzbekistan, analyzing their political systems, their relationships with the US and their initiatives against Islamic opposition groups. Chapter Seven provides a detailed examination and analysis of the most important radical Islamist group presently in and around the Ferghana Valley, the Hizb ut-Tahrir. Though most American readers will have never heard of it, they will be interested in seeing just how much this group already knows about them. Chapter Eight is the concluding chapter where the conundrum of trying to simultaneously fight terror in the short-term while spreading democracy in the long-term will be theoretically explained with the controversial conclusion that the United States is ostensibly building new bin Ladens.

Ultimately, this is not just a book about the inconsistencies of American foreign policy. It is also not just a book about Central Asian democracy flaws. These irregularities on our side and systemic abuses on their side are not simply detrimental to the development of functioning democracy in Central Asia, though truly that is so. These transgressions on both sides are sowing present-day seeds that will ultimately grow and flower into future radicalism. This radicalism, fueled by twisted Islamic interpretation and tinged with quasi-Marxist and anti-globalization rhetoric, will not be satisfied by the overthrow of corrupt and illiberal home regimes. The ultimate prize for these radicals would be the destruction of the 'true' cause of their turmoil – the United States. It is America that is seen as the initiator of Central Asian/Islamic despair and destitution because it is perceived to be the great purveyor in real terms of moral relativism and diplomatic convenience.

This book takes up the banner of Senator McCain's concerns. While he was correct in lamenting the 'choices' between chaotic democracy and authoritarian stability, he forgot about a third, even more dangerous choice: a violent Islamic radicalism that will have not only a cultural, linguistic and religious connection to the general population but will rest on a political foundation that is logical, analytically sound and horrendously damning to the United States. It will disseminate quotes like the one that began this chapter from Secretary of State Condoleeza Rice and ask, 'where was our dignity and justice as you drank tea with the Tajik government? Where was our economic development and prosperity as you sent your multinational oil and gas conglomerates to Uzbekistan? Where were your democratic values and human rights as you praised and congratulated the new Kyrgyz head of state?' McCain and others like him should always remember that *true* engagement with states pursuing democratic development can indeed result in chaotic democracy. That chaos, however, would not be for the long-term and would be simultaneously providing said states with the tools, support and ideas to persevere and emerge from their chaos stronger and more stable. It is our *actual* engagement as presently constituted that allows for authoritarian stability and it gives the impression that we may even perhaps prefer it to a true democratic experiment. This is unacceptable because it gives the third way, the radical Islamic way, the ultimate legitimacy.

This book takes the reader inside these governments, inside radical Islam and inside our own US foreign policy. It asks a question that is so incredibly difficult to answer that the United States government has been reluctant to even attempt to answer it. Quite simply, can you fight terror and simultaneously spread democracy? Are the two objectives compatible? This work shows definitively that we have been incorrectly fighting the former in the short-term and thus compromising our principles on the latter so profoundly in the long-term as to inflame our own future security problems. Most importantly, it reveals for the first time how compromising principles of democracy in areas already sensitive to Islamic revolutionary thought ultimately undermines the global war on terror in the most damning and significant way: we are, in the end, making the fight against terror not simply intractable but infinite in length and bottomless in its pool of future fighters who will come and try to kill us. In such a scenario that onus will not be on them, it will be on us.

The long-term answer to combating terrorism is indeed freedom, liberty, prosperity, dignity and justice. Societies strongly imbued with these characteristics will never be overwhelmed by terrorism. They may be attacked, but they will never be overcome. That is the best victory we can hope for. As long as we continue to compromise that long-term victory in the name of so-called short-term strategic gain we betray democracy in the name of combating terror. Again this matters because only real democracy defangs terror. If we continue to refuse to practice this fact while simultaneously preaching it as reality, we become our own source for a potentially fatal political sin.

Most world hegemons throughout history were never able to see their downfall. This was because they were constantly looking outward for oncoming threats, secure in the stability and superiority of their own societies. They were never prepared for the downfall from within. While it may be melodramatic to draw parallels at this point in time, it is nonetheless worth watching: America's downfall will not come at the hands of terrorists, but it may come because of the hypocrisy and contradictions that it created while fighting them. This book tries to bring this to light in the hope that knowledge leads to prevention. It is not yet too late to correct the flaws of our foreign policy. We have been saying the right principles and supporting the right program for quite awhile. What we need to do now is actually *act* in accordance with those principles and back up that program with real consequences when it is not adhered to. In the post-9/11 world, real, uncompromised, long-term democratic development as a remedy against terror must no longer be considered a secondary priority compared with short-term strategy and it especially must no longer be considered a haven only for idealists.

---

1 Speech by Secretary of State Condoleeza Rice at the Eurasian University, Astana, Kazakhstan, 13 October 2005.

2 Fareed Zakaria, *The Future of Freedom: Illiberal Democracy at Home and Abroad*, W.W.Norton & Co., New York, 2003.

3 Amy Chua, *World on Fire: How Exporting Free-Market Democracy Breeds Ethnic Hatred and Global Instability*, Bantam Doubleday Dell Publishing Group, 2004.

4 Noam Chomsky, *Hegemony or Survival: America's Quest for Global Dominance*, Henry Holt & Co., 2004.

5 Francis Fukuyama, *Statebuilding: Governance and World Order in the 21st Century*, Cornell University Press, 2004.

6 Natan Sharansky and Ron Dermer, *The Case for Democracy: The Power of Freedom to Overcome Tyranny and Terror*, Public Affairs, 2004.

7 Central Asia-Caucasus Analyst, SAIS, Johns Hopkins University, 'Uzbekistan Abolishes Media Censorship…', http://www.cacianalyst.org/view_article.php?articleid=1186, accessed 05 October 2005.

8 Central Asia-Caucasus Analyst, SAIS, Johns Hopkins University, 'Uzbek Human Rights Activist Sentenced', http://www.cacianalyst.org/view_article.php?articleid=775, accessed 10 October 2005.

9 Central Asia-Caucasus Analyst, SAIS, Johns Hopkins University, 'Uzbek President Granted Immunity for Life', http://www.cacianalyst.org/view_article.php?articleid=1412, accessed 17 June 2004.

10 Central Asia-Caucasus Analyst, SAIS, Johns Hopkins University, 'US Diplomat Raises Human Rights Concerns with Uzbekistan', http://www.cacianalyst.org/view_article.php?articleid=1786, accessed 19 September 2003.

11 Central Asia-Caucasus Analyst, SAIS, Johns Hopkins University, 'Uzbekistan Tightens Control Over Groups', http://www.cacianalyst.org/view_article.php?articleid=2089, accessed 12 February 2004.

12 Central Asia-Caucasus Analyst, SAIS, Johns Hopkins University, 'Tajik Opposition Party Protests Official Repression', http://www.cacianalyst.org/view_article.php?articleid=996, accessed 21 May 2002.

13 Central Asia-Caucasus Analyst, SAIS, Johns Hopkins University, 'Tajikistan's Rakhmonov Strengthens Long-term Grip on Power', http://www.cacianalyst.org/view_article.php?articleid=1545, accessed 30 June 2004.

14 Central Asia-Caucasus Analyst, SAIS, Johns Hopkins University, 'Tajik Opposition Party's Registration Rejected', http://www.cacianalyst.org/view_article.php?articleid=2268, accessed 24 July 2005.

[15] Central Asia-Caucasus Analyst, SAIS, Johns Hopkins University, 'Lawmaker Slams Marred Balloting in Kyrgyzstan, Tajikistan', http://www.cacianalyst.org/view_article.php?articleid=3138, accessed 23 September 2005.

[16] Central Asia-Caucasus Analyst, SAIS, Johns Hopkins University, 'Kyrgyz Forces Break up the Protest', http://www.cacianalyst.org/view_article.php?articleid=3429, accessed 18 August 2005.

[17] Central Asia-Caucasus Analyst, SAIS, Johns Hopkins University, 'Kyrgyz Authorities Arrest Ex-speaker of Parliament', http://www.cacianalyst.org/view_article.php?articleid=3433, accessed 19 August 2005.

[18] Central Asia-Caucasus Analyst, SAIS, Johns Hopkins University, 'Kyrgyz Deputy Suggests Year-long Moratorium on Protests', http://www.cacianalyst.org/view_article.php?articleid=3753, accessed 09 November 2005.

[19] Central Asia-Caucasus Analyst, SAIS, Johns Hopkins University, 'OSCE Pleased with Kyrgyz Presidential Ballot', http://www.cacianalyst.org/view_article.php?articleid=3521, accessed 15 November 2005.

[20] The United States Mission to the European Union, 'Assistant Secretary of State Jones on US Policy in Central Asia', http://www.useu.be/Terrorism/USResponse/Dec1401JonesUSPolicyCentralAsia.html, accessed 12 April 2004.

[21] Institute for War and Peace Reporting, 'Will US Policy Backfire in Central Asia', http://www.muslimuzbekistan.com/eng/ennews/2004/03/ennews30032004a.html, accessed 21 December 2004.

Chapter 2

Wonka Vision: Spreading Democracy or Seeding New Terror?

Over the past century, our nation has been called on to stand up for liberty – not only for our own but for that of others. We have answered that call many times. It is no different today... In each case, the enemies of freedom wish not only to dominate and kill, but to confuse and demoralize us and our allies. They cannot succeed unless we let them. We are freedom's force. And freedom, we know, is the true foundation of justice and peace.[1]

Kim R. Holmes,
Assistant Secretary for International
Organization Affairs

In general terms, official US foreign policy to the Central Asian region has always been three-fold: preventing the spread of terrorism, providing tools for political/economic reform and instituting the rule of law, and ensuring the development of energy reserves.[2] This chapter takes the US government's specific policy on the Ferghana Valley states, keeping in mind the larger, overall foreign policy of the US, and examines it from two analytical perspectives: how foreign policy is professed philosophically and how foreign policy is implemented financially.

Former Secretary of State Colin Powell declared, 'freedom, prosperity and peace are not separate principles, or separate policy goals. Each reinforces the other, so serving any one requires an integrated policy that serves all three'.[3] Despite these fine words, there is a strange disconnect when it comes to US foreign policy between the world our government sees and the world that actually exists on the ground. This is especially so in the three states who share the Ferghana Valley. In fact, US policy seems to suffer from some odd *Wonka vision* of democracy, where apparently a belief rules that as long as the right buzz-words are repeated over and over across diplomatic agencies reality will come to reflect the fantasy. As will be shown in this chapter, nothing could be further from the truth.

By comparing philosophy against implementation it is revealed whether policy maintains a level of consistency or contradiction. This type of analysis is essential in understanding the true state of national security: philosophically, policy always tends to be professed in the most grand and optimistic of terms. How that policy is ultimately financed and implemented, however, and how the recipients are ultimately impacted, is the true indication of whether or not policy produces results. Philosophy without empirical confirmation in policy circles quickly degrades into meaningless rhetoric. When it comes to the war on terror, rhetoric in place of results is playing a dangerous game with American security.

The United States Human Rights and Democracy Strategy

There is no loftier institution or holier chalice in American political philosophy than democracy. The Bureau for Democracy, Human Rights, and Labor (DRL), a governmental organization committed to supporting and promoting democracy programs throughout the world, says it most poetically by explaining that democracy is the one national interest that helps to secure all the others. Democratically governed nations are more likely to secure the peace, deter aggression, expand open markets, promote economic development, protect American citizens, combat international terrorism and crime, uphold human rights, and avoid humanitarian crises.[4] For the DRL, this is largely done through its Human Rights and Democracy Fund (HRDF), which was established during the Clinton Administration in 1998.

The DRL, hyperbole aside, represents the diplomatic frontline for American foreign policy when it comes to spreading democracy. The Bureau concisely and boldly expresses American priorities without equivocation:

- Assist newly formed democracies in implementing democratic principles
- Assist democracy advocates around the world to establish vibrant democracies in their own countries
- Identify and *denounce* (italics and bold mine) regimes that deny their citizens the right to choose their leaders in elections that are free, fair, and transparent.[5]

According to the DRL, the United States is pursuing a broad strategy of promoting respect for human rights that is both morally self-justifying and strategically beneficial for US security. The US is 'persuaded that regimes that violate the human rights of their own citizens are more likely to disrupt peace and security in their region and create a reservoir of ill will that can accrue to the detriment of the United States'.[6] This persuasion is backed up by a National Security Strategy that lists eight demands of human dignity:[7]

1 – The rule of law
2 – Limits on the absolute power of the state
3 – Freedom of speech
4 – Equal justice
5 – Respect for women
6 – Religious and ethnic tolerance
7 – Respect for private property
8 – Freedom of worship.

These eight demands of human dignity are not merely words on paper according to the DRL but is a policy stance matched by *action* in the way the US engages the world community. This engagement is a primary objective of US foreign policy, pushed forward by at least three goals that form the core of American strategy:

- Hold governments accountable to their obligations under universal human rights norms and international human rights instruments.
- Promote greater respect for human rights, including freedom from torture, freedom of expression, press freedom, women's rights, children's rights, and the protection of minorities.
- Promote the *rule of law, seek accountability, and change cultures of impunity.* (Italics mine).

There is nothing inherently wrong with the philosophy underlying US foreign policy. On the contrary, it is almost inarguable across all of its convictions. The problem is in the failure of implementation. This is not, however, a criticism of results or the lack thereof. Results are hard to gauge, as they may need a generation or two to truly take root and become evident. Indeed, it will be seen in this chapter how often US diplomats emphasize the intrinsic 'generational' quality of reform in the states of the Ferghana Valley. Unfortunately, this emphasis is a false promise: results can emerge only if policy is truly engaged and implemented. There will be no results, not immediate nor generational, if implementation is not genuine. While the US may be philosophically engaged with Central Asia it is not truly engaged in terms of productive implementation. It declares the right principles in regards to law, accountability and human dignity, but it chooses to ignore them if an ally in the war against terror makes a mockery of them. The US has not tried to change the culture of impunity and the present work simply wishes to reveal how unwise such games are when connected to terrorism.

Ironically, the DRL proudly touts the historical relevance of American foreign policy in impacting the globe for the spread of democracy at the end of the twentieth century. In 1974 there were but 30 countries who qualified as legitimate democracies. By 2005 that number had grown to 117. This, according to the DRL, will be one of the United States' greatest legacies. The only problem is that empirical reality provides more counter-evidence than confirmation.

The greatest and most immediate opportunity to increase the size of democracy's cradle occurred with the sudden dissolution of the Soviet Union in 1991. With its breakup the terrible fear of nuclear holocaust disappeared, at least via the traditional version of WWIII being waged between two superpowers with rival ideologies. In its place emerged 15 new independent states all favorable to democracy in one degree or another. Many in the United States argued to take advantage of this 'peace dividend' by supplying massive amounts of aid, assistance, and expertise. It was thought the conversion of the Soviet political space into free, independent, and democratic countries would bring in a new age of *Pax Americana*. To a certain extent this dream and vision still exists within American diplomatic circles only it has been tempered by more than a decade of stunted progress, hindered by post-Soviet governments overrun by corruption and economies that have stagnated.

Central Asia is a poster-child for this reality. It has been the slowest region of the former Soviet Union to change and convert. It has remained isolated politically and economically, dominated at home by regimes that are nearly absolute in their power and egomaniacal in their rationalizations. Before 9/11 the region was simply too insignificant, in comparison with Russia, to be a major front on DRL's agenda. After 9/11 the three states around the Ferghana Valley were a huge strategic asset in a larger foreign policy that had sights set on Afghanistan, Iraq and ultimately engagement in some form or another with Iran.

Despite this possible boon to Central Asia there has been no functioning implementation of democratic principles. There has been no real oversight to gauge whether vibrant democracies are being established. And there has certainly not been a process where regimes have been singled out and *denounced* for the fact that they have consistently denied their citizens the right to choose their leaders and engage their governments in peaceful opposition and open debate. These were the principles meant to power US policy. They were the core of its human rights and democracy strategy. But reality is wholly different as it shows US policy to be rather impotent. It is a *Wonka Vision* of democracy that will ultimately come back to haunt the US in the war against terror.

The Wonka Vision of Democracy in Central Asia

From September 11, 2001 to the present day consistent lip-service has been paid to the need for connecting the global war on terror with three state-building ideals: respecting democratic principles, developing civil society, and instituting the rule of law. Time and again across numerous governmental agencies and non-governmental organizations this connection has boomed so loud and clear that there seems to be a fantastic four of American diplomacy:

1 – Fight terror
2 – Respect democratic principles
3 – Develop civil society
4 – Institute the rule of law.

The United States says it is not interested in finding convenient partners who help it conduct the war against terror but who are not willing to produce true positive change within their own societies. The US says it is not willing to turn a blind eye to transgressions in the latter 3 goals mentioned above in exchange for assistance in the first. As former Secretary Powell said, each reinforces the other. This symmetry/symbiosis between fighting terror internationally while encouraging democratic development within allies domestically is clearly America's present philosophical foundation. Ostensibly, this foundation emerged as a consequence to the harsh lesson learned after the US disengaged from Afghanistan in the early

1990s: countries must not be left alone and allowed to become 'breeding grounds' for extremism and terrorism.[8] To prevent these destructive forces from emerging in Central Asia US diplomacy has supposedly *intensified* its efforts to help the countries of this area become 'stable, prosperous, and fully integrated into the world community'.[9]

Unfortunately, these claims appear to be hollow. Funding to the Ferghana Valley states did indeed consistently and significantly increase after 9/11. Yet very little on the ground has changed with this funding increase. While the Ferghana states have cooperated extensively in the war against terror (in some ways perhaps too enthusiastically inside their own borders) there has been very little progress in regards to the three key state-building objectives. In many cases, when it concerns democratic principles, civil society and rule of law, Tajikistan, Kyrgyzstan and Uzbekistan have been backsliding.

Despite this fact US diplomatic and governmental actors have continued to laud so-called 'improvements' while tacking on throwaway lines about more 'work to be done'. Is this permissiveness coincidental to the fact that all of the governments sharing the Ferghana Valley were generous in offering assistance after 9/11? Is this disparity between observation and reality at all tied to the fact that these governments quickly joined the coalition against terror and offered 'whatever was needed' to the US? Does Tajikistan's corrupt narco-economy or Uzbekistan's religious inquisition government gain patience and understanding from the United States because of true efforts to reform or because they were the first to offer the use of their facilities to coalition troops and both played central roles as staging areas for crucial early operations against Al Qaeda and the Taliban? Is it irrelevant that Kyrgyzstan was chosen to be the primary base for coalition air support and has also had the problems of its new revolution overlooked? Is it merely fascinating timing that the bombing campaign in Afghanistan began on October 7, 2001 *one hour* after Uzbekistan and the United States signed the Status of Forces Agreement that gave the US the legal right to carry out search and rescue missions from Uzbek sovereign territory?[10]

This divergence between philosophy and fact forms the basis of *Wonka Vision* democracy. With a nod to the children's fantasy classic, *Charlie and the Chocolate Factory*, *Wonka Vision* argues that it is OK to ignore the empirical reality of false democracy while professing continued admiration for the fantasy democracy supposedly emerging. It naively sacrifices long-term objectives in favor of short-term strategy. By pursuing such a strategy one of two accusations against the present administration is possible: either the US government is truly unaware of what is happening on the ground (unlikely) or it is being foolishly short-sighted in its positioning within Central Asia (likely). For now, reality shows that the United States says YES to fighting terror and developing energy reserves but it seems to be saying NO to helping democratic institutions truly emerge. The first accusation would be damaging enough to the long-term goals of the United States but it is hard to imagine that ignorance is the root of this problem. The second more likely accusation is devastating to US security interests as only real democratic development fights terror in the long-term stable manner we profess to seek.

Unfortunately, as long as the first two objectives of short-term counter-terrorism and energy development are secure, then the third objective of real democratic reform is apparently allowed to rot on the vine while the US, Kyrgyz, Tajik and Uzbek governments publicly proclaim 'acceptable progress'. If the United States truly wants to defeat Islamic terrorism it must seek to end *Wonka Vision* democracy. It cannot be satisfied with simply stating it maintains an 'intense dialogue' with such governments. It must cease providing governments with the rational justification for democratic deviation by emphasizing the need for a 'generational shift' in leadership.[11] How does such a shift or expansion come if American diplomacy is hopelessly empty and accommodating? It is rather fatally ironic that *Wonka Vision* democracy sows the seeds of new terror. The Ferghana Valley is but one of those fertile fields that we ourselves ignorantly cultivate.

'African Aid' Syndrome – US Assistance Programs and Money

Since the fall of the Soviet Union it was clear that US government funds in the form of assistance programs were meant to act as a stabilizing force and key element in the post-Soviet transition. Indeed, political and economic transformation, nuclear non-proliferation, aid in legal/judiciary reform, promotion of party development and free media have all been major technical points of departure for these USG-funded programs. Unfortunately, the same tortured diplomatic language emblematic of *Wonka Vision* democracy at State Department press conferences has also seeped into the official rationale used to explain USG financial assistance increases even though there are few if any significant results on the ground.

Ambassador Carlos E. Pascual, who was named coordinator of US Assistance to Europe and Eurasia in 2003 and subsequently elevated to coordinator for stabilization and reconstruction in 2004, used the same 'generational' rationalization when describing the success level of fiscal year 2003.

> Though the progress of this *generational transition* has varied tremendously among the countries of the region in fiscal year 2003, USG assistance programs continued to help promote good governance, strengthen civil society, independent media, the rule of law and human rights…Yet assistance has been used not only to build the security and law enforcement cooperation necessary for the struggle against global terror. It has also been deployed *to address the internal conditions that may lead, over time, to conflict and extremism* and the emergence of new 'failed states' which can become breeding grounds for terrorism.[12] (Italics mine).

The term *Wonka Vision* is not used flippantly. Simply no other term so accurately describes the prevailing tendency within the US government to either ignore reality or explain it away with a simple generational dismissal.

As coordinator of US assistance to Europe and Eurasia, Ambassador Pascual oversaw the development of regional and country strategies to promote market-oriented and democratic states and ensured that U.S. assistance reinforced American interests. He managed the allocation and implementation of approximately $1.1 billion in annual assistance. His new position, at the Office for Reconstruction and Stabilization, was meant to lead U.S. government planning, institutionalize US capacity, and help stabilize and reconstruct societies in transition from conflict or civil strife so that they could reach a sustainable path toward peace, democracy and a market economy.

These are all lofty, admirable goals well-expressed in powerful and moving mission statements. But in the end it is a lot of screaming into the wind and perhaps purposefully so: for while the United States is continually using the right language to frame the debate it is continually elevating financial expenditures despite the lack of results. Considering this is at present a Republican administration which has often spoken of the need to pinpoint foreign aid to 'documentable' results, it is confusing to understand how Tajikistan, Kyrgyzstan and Uzbekistan seem to be sliding by with a free pass. Confusing at least until a lens of cynicism is placed over policy so that references to democracy are deleted and mentions of security and anti-terrorism are left untouched.

A moment now to quell the hawks that argue conducting USG assistance programs along a strict axis of security, law and anti-terrorism is justifiable. This work is not saying it is fundamentally wrong to fund programs that exclusively consider short-term American national interests, though it does suggest that such strategy may not ultimately be our wisest course of action. The real problem with US funding since 9/11 has been a decided tendency to not only continue funding to questionable states but to increase funding to the Ferghana Valley states in areas strictly aimed at promoting democracy and civil society even though in reality they are simply being rewarded for security cooperation. Most disturbingly, this cooperation ultimately places our future security at risk.

United States Agency for International Development

The US seems to be suffering from what is akin to 'African Aid' syndrome: one of the fundamental problems in African democratic and economic development was an inexplicable tendency among international lending institutions to grant huge sums of financial assistance only to then not oversee its utilization or monitor its distribution. Inevitably, African governments quickly learned how to diplomatically say the right things to the international community while actively using funds in any way they pleased.

Thankfully it does not appear that Tajikistan, Kyrgyzstan and Uzbekistan are intentionally modeling themselves after the worst offenders from the African continent. But it does seem difficult to find justification for the financial increases considering these three states are nowhere near being functioning democracies nor do they show any real progress on the ground. How do funds get accounted for? What oversight mechanisms exist? Why are there no results timetables tied to the release of new funding? Accountability, oversight, timetables. These three things are essential if the US is sincere in wanting the states who share the Ferghana Valley to become actual democracies and join the global community. To be short, money as a reward for helping fight terror in the short-term is not enough if that money is used domestically in ways that does NOT respect democratic principles, does NOT develop civil society and does NOT institute the rule of law. As will be seen in the later Central Asian chapters, the United States creates its own vicious circle of terror when foreign policy is conducted in this manner.

Table 2.1 Fiscal Year 2002 Expended Funds – Non-Security Measures

US Agency International Development	Kyrgyzstan	Tajikistan	Uzbekistan
Democratic Reform	$5,130,000	$2,970,000	$2,050,000
Humanitarian Assistance	$540,000	$1,700,000	n/a
Social-Sector Reform	$6,920,000	$3,530,000	$160,000
Eurasia Foundation	$2,850,000	$1,100,000	$2,050,000

The US Agency for International Development (USAID) conducted the vast majority of its USG-funded programs bilaterally, closely interacting with local actors who were inevitably connected to and often dependent upon elite structures. USAID is supposedly committed to supporting the transition of authoritarian, centrally planned societies to participatory democracies with strong market-based economies.[13] Promoting regional cooperation and stability is crucial across Eurasia, a region critical to US national interests and the global war on terror, according to USAID. Written large, its regional programs in the Ferghana Valley were meant to promote policy reform, institutional development, broad-based citizen participation, and the rule of law.[14]

Nearly thirty million dollars was specifically spent in fiscal year 2002 on programs meant to directly impact the long-term democratization effort. These programs are not charitable donations or throwaway money. Democratization success fights terror best. This funding has increased every year since fiscal year 2002. Yet in all four areas documented above – democratic reform, humanitarian assistance, social-sector reform and the Eurasia Foundation, which itself runs programs meant to elevate individuals and societies through democratic and market transitions – there was little to no identifiable institutionalized progress. Worse, to the best knowledge of this author, there were *no real oversight mechanisms* to manage and monitor the distribution of these funds. Once distributed funds went figuratively into the political abyss with little to no recourse for the granting organization.

Freedom Support Act

Thomas C. Adams, who was the acting coordinator to Europe and Eurasia in 2003 before the designation of Ambassador Pascual, gave a perfect example of *Wonka vision* democracy when giving a statement to the House International Relations Committee on assistance to Eurasia in Washington DC:

> The picture on democratic reform is decidedly less rosy. Noticeable backsliding has occurred in recent years in some countries. While I do not diminish these problems I believe that we have achieved tremendous accomplishments in this sphere as well over the past decade and have sown the seeds for future positive change.[15]

Again, despite these statements, the only real definitive accomplishments made in Central Asia since 9/11 have been in the military and security spheres where these states have allowed US forces almost unfettered access to their sovereign territory. Success is always referred to in terms of positive democratic change even though there are no real examples to support it. Coordinator Adams went on to describe one of the largest assistance programs ever created in the United States:

> One of our best mechanisms to foster fundamental change is our program of professional and educational exchanges. Freedom Support Act exchange programs have brought slightly over 100,000 business people, journalists, students and professors to the United States to see for themselves how free people prosper and deal with challenges common to all countries…FSA funds have supported the development and strengthening of civil society organizations and non-governmental organizations…[and has] also supported the development of political parties, helped to strengthen legislatures as a counter-weight to powerful executive branches of government, and worked to create or bolster judicial independence.[16]

The fact of the matter is that none of these things have happened across Central Asia let alone in the three countries sharing the Ferghana Valley. Political parties are still nondescript and emasculated. Legislatures remain rubber-stamp agencies of the hyper-powerful executive branches. The mere idea of an independent judiciary is inconceivable at this point in time in the region. And yet time and again the highest diplomatic levels speak of progress happening all over the area and largely thanks to the aid and assistance of the United States. More fascinating (frustrating?) still was how Adams mentioned the three Ferghana states by name in his presentation:

> The largest changes in the FSA budget request are related to our country's intensified engagement with Central Asia as a result of the September 11 attacks on our country and the global war on terrorism. Significant increases for Uzbekistan, Tajikistan, and the Kyrgyz Republic will support economic development in conflict-prone areas, effects to open political space by strengthening democratic institutions and grassroots organizations.[17]

Table 2.2 Freedom Support Act Funds Expended FY 2001–2003

FSA Funds Expended	Kyrgyzstan	Tajikistan	Uzbekistan
FY 2001	$33,390,000	$16,800,000	$25,840,000
FY 2002	$40,000,000	$32,330,000	$39,660,000
FY 2003	$43,600,000	$35,000,000	$42,000,000

It is interesting to note that this work will document in its Central Asian chapters how the only significant grassroots organizations that have emerged and begun to form real opposition to the ruling elites since 9/11, despite FSA budget increases, have been *radical Islamic fundamentalist groups*. Clearly this was not the democratic opposition or grassroots activism the United States had in mind when it consistently elevated its financial engagement with the Ferghana Valley states. But it is exactly this kind of opposition it should have anticipated when considering its analysis of progress and estimation of its own 'democratic engagement'.

Human Rights and Democracy Fund

DRL's Human Rights and Democracy Fund (HRDF) is used to fulfill the Bureau's mandate to monitor and promote human rights and democracy worldwide. According to its own mission statement, the HRDF supports 'innovative programming designed to uphold democratic principles, support democratic institutions, promote human rights, and build civil society in countries and regions of the world that are geo-strategically critical to the US.[18] It has since 9/11 occupied a major foreign policy niche in financial terms, receiving enormous budget increases and gaining incredible responsibility the world over for initiating programs like the ones mentioned above.

By criteria it is not surprising that Tajikistan, Kyrgyzstan and Uzbekistan is a major focus of the HRDF. Every aspect of the Bureau's vetting process to determine important targets is satisfied by the three states of the Ferghana Valley. There can be no doubt once the war against terror began that Central Asia would elevate in geo-strategic importance. In addition, the supposed 'philosophical commitment' to democratic principles, civil society and human rights since 9/11 had no better contemporary focus than the Ferghana three in that they epitomized the type of state and location the US government would like to engage. Sure enough, the HRDF even singled out the Central Asian region on its home page, emphasizing that USG funds in Central Asia had a 'renewed commitment to promoting human rights both diplomatically and programmatically'.[19] A casual glance at the level of USG funding over the last four years to the HRDF indicates the increasing importance this fund was meant to have when it came to on-the-ground implementation across these geo-strategic regions of importance.

Table 2.3 Fiscal Year Budget Increases for the HRDF

Fiscal Year	USG Funds
2000	$8,969,985
2001	$13,421,000
2002	$13,000,000
2003	$36,448,000
2004	$43,740,000

So what kinds of 'innovative programming' were seen in the three Ferghana Valley states? What sort of on-the-ground effects emerged with this nearly 400% percent increase in HRDF funding since 2000? In 2000–2001 the two largest programs involved 'capacity-building for human rights advocacy NGOs' in Uzbekistan and funding 'get-out-the-vote' NGO activities in Kyrgyzstan. Ironically, it was during this time period that both Uzbekistan and Kyrgyzstan earned a reputation for government harassment and investigation into foreign-run organizations. This severely curtailed the operational environment for NGOs in these countries.[20]

In 2001–2002 the NGO capacity-building budget in Uzbekistan was increased 500% and was combined with an Uzbek 'rule of law' program funded for half a million dollars. Apparently the 'get-out-the-vote' NGO program in Kyrgyzstan was so successful that it earned its own elevation in status by being transferred into a new 'capacity-building' Kyrgyz program. Over half a million dollars went into funding this program as well. Another half million went to a program called 'Citizens' Rights in Ferghana Valley', which went to creating an advocacy network throughout the area. In all three cases it seems apparent that over three million dollars were simply washed away and ultimately wasted as there was no real follow-up in how these funds were truly utilized.

In 2002–2003 all previous programs were maintained while 'political party development' emerged to take center-stage on the HRDF agenda. An additional one million dollars total went to financing the development of political parties in both Uzbekistan and Kyrgyzstan. In its own programming the HRDF exalted such programs' relevance by claiming the need to 'encourage constructive dialogue' between the government and opposition groups.[21] However, the most vehement opposition to government across the Ferghana Valley has been the emergence of Islamic political parties and Muslim organizations that almost unequivocally stand for the forced removal of the present government. I would not consider this as an example of encouraging constructive dialogue.

In 2003–2004 the largest expansion of HRDF programs covered the Ferghana Valley, only confirming the supposition that Central Asia was indeed a major focus of interest for its four-fold increase in overall budget. Political party development in Uzbekistan and Kyrgyzstan increased another half million dollars. Interestingly, new programs seemed to shift off of democracy building and move toward humanitarian issues, as Uzbekistan received over one million dollars to institute such programs as the 'Civic Coalition Against Torture' and 'Protection of Human Rights'. Tajikistan also received the Human Rights Protection program while Central Asia as a regional whole received funding for a 'Civic Bridges' program meant to create a regional, interactive network of citizens which would promote civic participation and advocacy programs.[22]

It is fascinating to note that as the funding for HRDF programs in Central Asia skyrocketed, program content gradually shifted from democratic state-building agendas to international law and human rights initiatives. This was not a coincidence but confirmation of the utter lack of success in developing opposition political parties that engaged in constructive dialogue and created civic participation networks across the region. Why, four years into this funding largess, was it necessary to begin programs aimed at stopping torture and protecting human rights? It seems logical to suggest that programs aimed at civil society and democracy, if they were having any sort of impact whatsoever, would at least have made inroads against such barbarity as torture. Instead, as US government programs became more embedded in the region, there seemed to be regression on the ground. This regression flies in the face of American diplomats like Assistant Secretary of State Elizabeth Jones, who lauded all efforts as being 'every bit as important as our security assistance in dealing with the *long-term root causes of terrorism*'.[23] Closer inspection across the board of American governmental agencies finds this same illusory satisfaction, inexplicably proclaiming Central Asian policy as one that was deep, sustained and coordinated.[24]

Hitting Your Talking Points: Selling US Policy on the Hill and at the Hague

House of Representatives International Relations Committee

If looking for a smoking gun or seminal moment in the selling of US foreign policy on Central Asia, then ironically one of the strongest contenders came on Capitol Hill before 9/11 actually took place. On 6 June 2001 US policy on Central Asia was discussed before the House of Representatives, specifically before its Subcommittee on the Middle East and South Asia. The reason I choose this moment as the defining one is simple: Central Asia's strategic relevance only became more intensified after the World Trade Towers collapsed. Any hope at all of policy toward Central Asia being held to a high standard of introspection and oversight would had to have been in place before the security crisis known as the global war against terror had begun. This opportunity to build such a foundation did indeed emerge in summer 2001 but was subsequently passed up.

Overall, the members of the House International Relations Committee, chaired by Representative Henry Hyde of Illinois, displayed an expected mixture of facts and falsehoods when discussing Central Asia. Representative from New York Benjamin Gilman admitted to 'serious human rights problems and extremist movements influenced by Afghanistan', but also lauded the region for its 'enormous energy export potential that could ease current US energy problems and for its governments being secular'.[25] Gilman's statement is fascinating for it being both prescient and wrong at the same time.

Central Asia's energy reserves, though significant, face severe obstacles of climate, geography and technology. At present there is some doubt as to the actual extent of the energy reserve estimates and that perhaps estimate inflation may have been done in order to secure more international aid. Regardless, Gilman accurately framed the debate: he showed concern at the political repression but wrapped up his analysis with praise for the economic potential. The lamentable fact was how ignorant many of his observations were toward the actual doings of the Central Asian governments: calling them secular when in fact they manipulate Islam for their own purposes and are ruthless in the harassment of Islamic organizations is naïve at best. At worst, it is purposefully ignoring the preconditions capable of creating Islamic terrorism. As we shall see below, Rep. Gilman was not alone.

The main figure to appear before the House subcommittee that day was Clifford Bond, a career diplomat who at the time was the Acting Principal Deputy to the Special Advisor for the New Independent States. He was called onto the floor as the administration's official expert witness on Central Asian policy. His opening statement was pregnant with the hidden problems brought to light in this work:

> The overarching and the long-term goal of US policy in Central Asia is to see these states develop into stable, free-market democracies, which can serve as a bulwark against the spread of potential instability and conflict in the region. This broader goal serves three core strategies or interests of the United States: regional security, political and economic reform and energy development.

Here, before 9/11, before the global war on terror, the three core values in US policy are made explicit. Since this performance on Capitol Hill it has become clear that the first and third core interests – regional security and energy development – exploded and overwhelmed the more important long-term second interest – political and economic reform. Indeed, Bond would obliviously go on to openly discuss this fundamental divergence that would come to epitomize US foreign policy:

> All the states of Central Asia have indicated that they welcome security cooperation with the United States. They see our engagement in the region as an additional element of stability as they seek to balance their relations with more powerful neighbors in the region…Only by empowering their citizens through

democratization and economic reform can these governments ensure lasting popular support and stability. This is an *integral part* of our message to the governments of Central Asia' (italics mine).[26]

It is ironic how before 9/11 the United States viewed itself as the security blanket necessary for Central Asia so that governments like Kyrgyzstan, Tajikistan and Uzbekistan would be able to 'fend off' bigger, more powerful rivals like China, Iran and Russia. After 9/11 the security relationship was flipped upside down, as it was the United States that needed Central Asia when it came to its *own* immediate national security. This flip was what most likely empowered the Ferghana Valley states to essentially ignore and dismiss Bond's second priority of democratic and economic reform. Before 9/11 the leverage was on the side of the United States. After 9/11 the leverage shifted decidedly to the side of the Ferghana Valley states. It seems that America decided in favor of making its short-term security priorities supreme. To this day we do not publicly acknowledge how this has made it difficult to be principled on democratic development.

It is one thing for Representative Gilman to make incorrect statements about a Central Asian state. He was not a recognized expert and did not serve the government in such a capacity. It is cause for serious concern however when false and disturbingly vacuous statements are made by the official administration point-man on Central Asia. Such was the case when Bond discussed Kyrgyzstan's perplexing backsliding when it supposedly had been the best in the region in protecting human rights.

> It is difficult to understand or explain President Akaev's motives over the past two or more years, but we have seen a turn toward repression since 1999 and the onset of a set of parliamentary and then presidential elections in the year 2000. In the lead up to those elections and as the opposition became more vociferous and more organized, we saw harassment of opposition figures…we saw the exclusion in the presidential elections themselves of principal opponents to President Akayaev. We saw that expand beyond activities directed against individuals to the suppression of the free media, to the harassment of NGOs.[27]

Though Bond was perplexed, the fact was that the Kyrgyzstan administration had no real interest in 'constructive dialogue' or the development of such institutions as civil society and political parties as these things that could ultimately be a threat to its vice-like grip on power. As Bond said, just as alternative voices became louder and more organized the central government cracked down. This would not be confusing to an objective observer absent a vested interest. Remember that this came before 9/11 so the convenient lack of understanding came with the US only thinking of Central Asia as a counter to China and Russia and as a potential source down the road to greater non-Arab energy reserves. How much more was the United States government willing to overlook democratic backsliding and human rights violations when 9/11 happened and Central Asia had become important in the name of critical short-term security concerns? This is not rocket science, just

Machiavellianism. Indeed, Bond commented on the 'tough stance' the US was willing to take against this regression, saying that, 'we are engaged in a dialogue with the government. We have made some progress in modest ways to get them to respect human rights, but frankly they need to do a lot more, and we need to continue to talk to them'.[28] Considering its record in the years to come, it appears the central government of Kyrgyzstan was not greatly concerned by such 'harsh language' from the United States.

Even more disturbing was when Bond discussed radical Islamism within Central Asia:

> Let me say, first of all, that Central Asia is not fertile ground for Islamic fundamentalism. The Islamic tradition in these countries is a very tolerant one. The Taliban's variety of Islam has little attraction to the great mass of people who accept the idea of a secular state…we do not see Islamic fundamentalism right now as a threat to the states of Central Asia.[29]

In the most direct terms it was not a threat because the 'secular' governments of the Ferghana Valley were willing to do anything to co-opt, undermine and oppress Islamic political groups. But this type of reasoning was dangerously myopic, ignoring as it did the fact that it contradicted our policy aimed to encourage democratization and civil society opposition. It was also myopic because across the globe there were many cases of Islamic opposition groups being forced underground by oppressive governments. The one universal result to emerge from such maneuvers was a radicalization of said Islamic groups. The recipe to create such a potentiality was already on the ground in the Ferghana Valley because of the repressive and violent responses of the local executive governments. Bond was conveniently and inconceivably papering over the danger by saying Central Asian society practiced a more tolerant Islamic tradition.

The fact of the matter is that most of the world practices a more tolerant Islamic tradition. The threat in the global war against terror is not the fear of more STATES transforming into radical Islamic theocracies. There is no worry of 'Irans' mushrooming all over the globe. What should be the true fear undermining the global war against terror is repressive 'secular' governments that drive legitimate Islamic opposition underground and thus forces them toward radicalization. Central Asian society as a whole does not need to be radical or practice a non-tolerant version of Islam for it to be a breeding ground of radical, non-tolerant Islamic groups. This is the point Bond missed and it is the point that continues to be missed in US foreign policy today.

To show how this myopia was both contagious and relative to context, the dialogue between Principal Deputy Bond and Representative Dana Rohrobacher was most fascinating and, ultimately, disappointing and dangerous to long-term US and global interests.

> ROHROBACHER: You know, as much talk as there has been about the 'Silk Road' and about Central Asia, there has been very little done in Central Asia

by the United States government…because of that, for these last ten years what was a tremendous opportunity for expanding the democratic system into the region and opening up the region economically, I am afraid that by all of the criteria and judging how far we have come, we have not made very much progress.

…Tajikistan and Uzbekistan, especially Uzbekistan, that had great, great promise and very sophisticated leadership, instead of permitting freedom of elections and freedom of the press and having the United States lead these countries into a more democratic situation, we have seen no real lasting democratic reform in those countries…Instead, we have dictatorships which *drive the young people right into the hands of extremists, of Muslim extremists.* (Italics mine).

BOND: I do not disagree with your analysis about the state of democracy in Central Asia but…I think our assistance has had some successes. I think we have seen elements of a civil society develop in many of these countries…The US government has been instrumental in promoting the development of those, and hopefully those seeds will develop into democracy over time.

ROHROBACHER: I applaud the positive things you have done obviously. Our State Department has been trying. It just would seem to me that we should have been a lot tougher in demanding political reform…we end up giving economic concessions to some of the Central Asian countries, but not demanding political reform with those economic concessions. That I think *exemplifies the type of strategy that leads to no political reform and people taking it for granted that we do not really care about that political reform.* (Italics mine).

BOND: In terms of elite opinion, I have found them to be very interested in seeing the United States engage commercially because they think we bring the best technology and excellent management. They also want to see us engage politically and on security matters because, as I said before, they would like to see us balance some of their larger neighbors.

…[Kyrgyzstan] had a very vibrant media – it still does to some extent…it has political opposition figures who have organized political parties…it is a much more open political system…

…[Tajikistan] has more NGOs, more political opposition. There are actually Islamic fundamentalist opposition members in the Tajik parliament and more media, more free media…

> …Again, I said this before, and I do not want to sound like a broken record, but Uzbekistan has tremendous potential in the region if it would get its economic act together…
>
> ROHROBACHER: I am afraid that I do not see the light, even a glimmer that my colleague sees…There will not be economic reform in Central Asia until there is democratic reform in Central Asia. This is not what businessmen keep telling us about China: just invest and interact economically, and there is going to be democratic reform as a byproduct of economic progress. It is just the opposite.
>
> …You cannot expect people to invest in a dictatorship that has no free press and has no opposition parties. That is what keeps a country honest. That is what keeps a country honest enough for foreign investors to come in because they know they are not going to be asked for bribes or be intimidated by the corrupt officials in that country.
>
> …Of course, it is always easy to look for that glimmer, and maybe I hope that those people are looking for a more optimistic analysis of Central Asia. I am pretty pessimistic myself, but that does not mean that we cannot get there and work with those people and turn it around and try to make things better.[30]

The proceeding chapters on Central Asia will show how Principal Deputy Bond's specific comments were in fact incorrect. Ultimately, the progress and hope administration diplomats proclaimed only grew more false after 9/11, as our desire for democratic reform was replaced by our need for immediate security cooperation. Indeed, the problem with Representative Rohrobacher's comments was that he hit all the right points but then shunted them aside after the horror of 9/11. Just as Rohrobacher should have been screaming his criticism the loudest, he jumped on the security bandwagon and became silent.

Interestingly, Rep. Rohrobacher's colleague, Joseph Pitts from Pennsylvania, closed the subcommittee's meeting with a prepared statement that summed up the oncoming flaws of the 9/11 Central Asian shift:

> Economic prosperity, the growth of democracy, and the establishment of the rule of law in the Central Asian states is essential for US national security…unfortunately, US foreign policy towards the region emphasizes a stand back and watch approach…Blatant abuses of human rights have increased. Freedom of religion is nonexistent in some countries and getting worse in others. Terrorist cells of Islamic fundamentalists run free through many countries. The list goes on…But it is not too late. The US can still affect positive change in the region by engaging these countries. We must let them know that the US is not going to turn a blind eye to the region…If we engage them we can promote respect for human rights, and increase their understanding and acceptance of democracy and the rule of law.[31]

Poetic and prophetic, condescension notwithstanding, it was Representative Pitts who turned a blind eye, as did the rest of Congress, after 9/11. This was most unfortunate because only the long-term resolutions provided by democratic reform fully destroy the desperation that creates Islamic radicalism. The anger, shock, disgust and bloodlust unleashed by the indignity and tragedy of 9/11 clouded the judgment of the few lawmakers who appeared to have a proper grasp on the situation in Central Asia before the Twin Towers collapsed. The emotions are understandable but there should be no room for emotions in foreign policy if they compromise national interest and endanger long-term security.

Organization for Security and Cooperation in Europe

In general terms, US participation in the OSCE advances America's national interest in promoting democracy and strengthening respect for human rights. The OSCE therefore has a role to play in helping to win the global war against terror.[32] In a 2004 report to Congress, 'Increased Attention to Central Asia', it was obvious that little had changed since that day on the Hill in 2001. In fact, all of the contradictory hallmarks of US foreign policy that were there three years ago had apparently only become codified into talking points the world over. The piece was eerily reminiscent of Bond's talking points and indicative of a policy that proclaimed much philosophically but demanded little to nothing empirically.

> US goals in Central Asia are to develop democratic, market-oriented states that are more fully integrated into the Eurasian community, to promote regional stability, and to strengthen a long-term coalition against terrorism that respects human rights. To advance these goals through the OSCE, the United States will support new programs in the economic and political-military dimension in Central Asia…the *most promising* programs in Central Asia in support of US goals are those focused on trans-national issues, such as terrorism, trafficking, border security, police training, and judicial reform. *The Central Asians have shown willingness to cooperate on these issues.* (italics mine).[33]

If there was any difference between the two time periods it was simply the sad observation how over the course of three years and well into the global war against terror the stress on democratic reform and civil society had become but a throwaway line prefacing policy statements. The democratic dimension for accomplishing US objectives in the region had now been connected grammatically to the military through a timely placed hyphen. The most promising programs, the ones with which Central Asian governments were the most cooperative, were now strictly trans-national security programs. I am sure the irony and significance was not lost on the Ferghana Valley regimes that while the US officially lauded the OSCE for its work on 'anti-torture programs and trying to strengthen the freedom of the media', the US government was simultaneously creating an elaborate system of 'black prisons' to hold suspects in the war against terror and kept this system entirely secret from American media sources for at least 2 years.

The Moscow Mechanism

The Moscow Mechanism, a rarely used tool that gave the OSCE power to appoint a Rapporteur to investigate and ostensibly resolve serious international and interregional problems, was invoked in 2003 against Turkmenistan. The initial stimulus to invoke the mechanism was the manner in which Turkmenistan's security services conducted investigations into the attempted assassination of President Niyazov in fall 2002. Reports on the brutality and arbitrariness of those investigations prompted the United Nations and other international organizations to condemn the Niyazov government, thus pushing the OSCE to trigger the Moscow Mechanism. Once triggered, the Rapporteur's conclusions included a well-regarded set of recommendations for both the Turkmen authorities and the international community.[34]

Ultimately, the OSCE was unable to compel Turkmenistan to cooperate. The Turkmen government refused to issue even a visa for the Rapporteur and refused to conduct its own internal investigation honoring the procedures of the Moscow Mechanism. Ultimately, Turkmen authorities rejected the Rapporteur's final report and refused to implement its final recommendations.[35] This episode essentially destroyed the confidence of the OSCE that the Moscow Mechanism was an independent and effective international tool of investigation.

Even more significant was the relative silence by the US. Turkmenistan was arguably the worst state in Central Asia from a democracy perspective, representing a cult of personality that rivaled if not even surpassed North Korea. Did the United States government assume no lesson would be learned by the other states of Central Asia, especially the Ferghana three? Turkmenistan's ability to de facto emasculate what was considered one of the OSCE's more powerful diplomatic weapons set a scary and disturbing precedent: what were the chances that Kyrgyzstan, Tajikistan and Uzbekistan thought they would face serious repercussions if they ignored international criticisms when Turkmenistan, which had always rebuffed American overtures to assist in the terror war, did not face sanction or punishment? The Ferghana three actively cooperated in the war against terror. The lesson learned was clear: our policy mattered little. The empirical reality said it all. Security now mattered. Democracy later didn't.

Conclusion: The Dangers of Divergence in Diplomacy

The present analysis of the development of US foreign policy in Central Asia reveals a dangerous divergence that carries potentially deadly consequences. While lip-service is continually paid to the ideals and power of democratic freedom this verbal commitment is not backed up with results in diplomatic action or programmatic implementation. In fact the careful examination provided in this chapter revealed a very subtle shift since 9/11 in the type of programs funded in Central Asia – they focused less on democratic institutions and civil society and more on trans-national security issues of dubious achievement.

This problem is nothing more than the consequences to emerge when you have a divergence between the philosophical framing of a state's foreign policy and the empirical reality of its program implementation. There is no justification for this *Wonka Vision* of democracy. Time and again some of the most important political figures in the United States have walked to the podium, both here at home and abroad before foreign audiences, and declared a self-evident and crucial truth: democracy matters not just for freedom but for security.

Secretary of State Condoleeza Rice, on a trip through Central Asia in fall 2005, addressed this very issue with powerful, ringing words:

> We understand that the path to democracy is long and imperfect and different for every country. But make no mistake: the principles of democracy are unanimously desired and universally deserved. Here in Central Asia there is much to do but there are reasons for optimism…We all live in a world in which there are security concerns. We all live in a world in which terrorism is a threat…[but] you have to have democracy because *democracy is, in fact, the answer to terrorism and to extremism*. (Italics mine).[36]

Though powerful and ringing, these words ring hollow. Our philosophy rings hollow. The implementation of our programs does not reveal a sincerity of commitment. When President Bush spoke in 2003 at the National Endowment for Democracy he passionately declared that 'we know that liberty if not defended can be lost. By definition the success of freedom rests upon the choices and the courage of free peoples and upon their willingness to sacrifice'.[37] What this chapter has tried to reveal is how America, in its fight to supposedly preserve its own liberty today, may be turning a blind eye to how our 'allies' in this fight are undermining the liberty of their own peoples. If that remains true, and we are giving no real reasons to make it not true, then the sacrifices made in the future by the citizens of those countries may be sacrifices that come in the form of suicide bombers at our doorstep. You reap what you sow in foreign policy and at the moment we are cultivating a bitter future harvest.

---

1 Speech by Assistant Secretary for International Orgnization Affairs, Kim R. Holmes, at the Heritage Foundation's Annual Board Meeting, New Orleans, Louisiana, 17 April 2004.

2 The United States Mission to the European Union, 'Assistant Secretary of State Jones on US Policy in Central Asia', http://www.useu.be/Terrorism/USResponse/Dec1401JonesUSPolicyCentralAsia.html, accessed 12 January 2004.

3 The United States Mission to the European Unicn, 'Assistant Secretary of State Jones on US Policy in Central Asia', http://www.useu.be/Terrorism/USResponse/Dec1401JonesUSPolicyCentralAsia.html, accessed 12 January 2004.

4 The Bureau of Democracy, Human Rights, and Labor (DRL), 'Mission Statement', http://www.state.gov/g/drl/democ/, accessed 15 March 2005.

5 The Bureau of Democracy, Human Rights, and Labor (DRL), 'Mission Statement', http://www.state.gov/g/drl/democ/, accessed 15 March 2005.

6 The Bureau of Democracy, Human Rights, and Labor (DRL), 'US Human Rights and Democracy Strategy', http://www.state.gov/g/drl/democ/, accessed 15 March 2005.

7 The Bureau of Democracy, Human Rights, and Labor (DRL), 'US Human Rights and Democracy Strategy', http://www.state.gov/g/drl/democ/, accessed 15 March 2005.

8 Bureau of European and Eurasian Affairs, 'Frequently Asked Questions about US Policy in Central Asia', http://www.state.gov/p/eur/rls/fs/15562.htm, accessed 28 July 2004.

9 Bureau of European and Eurasian Affairs, 'Frequently Asked Questions about US Policy in Central Asia', http://www.state.gov/p/eur/rls/fs/15562.htm, accessed 28 July 2004.

10 Bureau of European and Eurasian Affairs, 'Frequently Asked Questions about US Policy in Central Asia', http://www.state.gov/p/eur/rls/fs/15562.htm, accessed 28 July 2004.

11 Bureau of European and Eurasian Affairs, 'Frequently Asked Questions about US Policy in Central Asia', http://www.state.gov/p/eur/rls/fs/15562.htm, accessed 28 July 2004.

12 Bureau of European and Eurasian Affairs, 'US Government Assistance to and Cooperative Activities with Eurasia', http://www.state.gov/p/eur/rls/rpt/37645.htm, accessed 27 August 2004.

13 Bureau of European and Eurasian Affairs, 'US Government Assistance Report 2004', http://www.state.gov/p/eur/rls/rpt/37677.htm, accessed 11 February 2005.

14 Bureau of European and Eurasian Affairs, 'US Government Assistance Report 2004', http://www.state.gov/p/eur/rls/rpt/37677.htm, accessed 11 February 2005.

15 Statement to the House International Relations Committee, 'US Assistance Programs in Europe: An Assessment', http://www.state.gov/p/eur/rls/rm/2003/19203.htm, accessed 24 March 2005.

16 Statement to the House International Relations Committee, 'US Assistance Programs in Europe: An Assessment', http://www.state.gov/p/eur/rls/rm/2003/19203.htm, accessed 24 March 2005.

17 Statement to the House International Relations Committee, 'US Assistance Programs in Europe: An Assessment', http://www.state.gov/p/eur/rls/rm/2003/19203.htm, accessed 24 March 2005.

18 Human Rights and Democracy Fund, 'Mission Statement', http://www.state.gov/g/drl/c7607.htm, accessed 13 October 2005.

19 Human Rights and Democracy Fund, 'Mission Statement', http://www.state.gov/g/drl/c7607.htm, accessed 13 October 2005.

20 Bureau of Democracy, Human Rights and Labor, 'Fact Sheets', Department of State, United States of America, 1 June 2004.

21 Bureau of Democracy, Human Rights and Labor, 'Fact Sheets', Department of State, United States of America, 1 June 2004.

22 Bureau of Democracy, Human Rights and Labor, 'Fact Sheets', Department of State, United States of America, 1 June 2004.

[23] The United States Mission to the European Union, 'Assistant Secretary of State Jones on US Policy in Central Asia', http://www.useu.be/Terrorism/USResponse/Dec1401JonesUSPolicyCentralAsia.html, accessed 12 January 2004.

[24] The United States Mission to the European Union, 'Assistant Secretary of State Jones on US Policy in Central Asia', http://www.useu.be/Terrorism/USResponse/Dec1401JonesUSPolicyCentralAsia.html, accessed 12 January 2004.

[25] US House of Representatives, Committee on International Relations, Subcommittee on the Middle East and South Asia, 'US Policy on Central Asia', http://commdocs.house.gov/committees/intlrel/hfa72975.000/hfa72975_0f.htm, accessed 12 June 2003.

[26] US House of Representatives, Committee on International Relations, Subcommittee on the Middle East and South Asia, 'US Policy on Central Asia', http://commdocs.house.gov/committees/intlrel/hfa72975.000/hfa72975_0f.htm, accessed 12 June 2003.

[27] US House of Representatives, Committee on International Relations, Subcommittee on the Middle East and South Asia, 'US Policy on Central Asia', http://commdocs.house.gov/committees/intlrel/hfa72975.000/hfa72975_0f.htm, accessed 12 June 2003.

[28] US House of Representatives, Committee on International Relations, Subcommittee on the Middle East and South Asia, 'US Policy on Central Asia', http://commdocs.house.gov/committees/intlrel/hfa72975.000/hfa72975_0f.htm, accessed 12 June 2003.

[29] US House of Representatives, Committee on International Relations, Subcommittee on the Middle East and South Asia, 'US Policy on Central Asia', http://commdocs.house.gov/committees/intlrel/hfa72975.000/hfa72975_0f.htm, accessed 12 June 2003.

[30] US House of Representatives, Committee on International Relations, Subcommittee on the Middle East and South Asia, 'US Policy on Central Asia', http://commdocs.house.gov/committees/intlrel/hfa72975.000/hfa72975_0f.htm, accessed 12 June 2003.

[31] US House of Representatives, Committee on International Relations, Subcommittee on the Middle East and South Asia, 'US Policy on Central Asia', http://commdocs.house.gov/committees/intlrel/hfa72975.000/hfa72975_0f.htm, accessed 12 June 2003.

[32] Bureau of European and Eurasian Affairs, 'US Policy and the Organization for Security Cooperation in Europe', http://www.state.gov/p/eur/rls/rpt/2004/31935.htm, accessed 31 October 2005.

[33] Bureau of European and Eurasian Affairs, 'US Policy and the Organization for Security Cooperation in Europe', http://www.state.gov/p/eur/rls/rpt/2004/31935.htm, accessed 31 October 2005.

[34] Bureau of European and Eurasian Affairs, 'US Policy and the Organization for Security Cooperation in Europe', http://www.state.gov/p/eur/rls/rpt/2004/31935.htm, accessed 31 October 2005.

[35] Bureau of European and Eurasian Affairs, 'US Policy and the Organization for Security Cooperation in Europe', http://www.state.gov/p/eur/rls/rpt/2004/31935.htm, accessed 31 October 2005.

[36] Speech by Secretary of State Condoleeza Rice at the Eurasian National University, Astana, Kazakhstan, 13 October 2005.

[37] Speech by President George W. Bush at the 20th anniversary of the National Endowment for Democracy, Washington DC, November 2003.

Chapter 3

Central Asia and The Ferghana Valley: A No Man's Land for Terror

The Ferghana Valley is an area quite rich in history that remains a controversial subject in academic circles. Today it is rather easy to find diametrically opposed viewpoints regarding the valley. On the one hand its relative stability and lack of overt violence is praised. On the other its precarious position as a tinderbox for Islamic fire is always feared. The present chapter will acquaint the reader with some of the valley's relevant history and analyze the divergent possibilities regarding its political future. The tinderbox view of Ferghana will be upheld and proven. The dismissal of Ferghana as a critical nexus of Islamic opposition is too often based upon a foolish standard that seeks to ascertain whether Islamic radicalism can actually succeed in overthrowing a state. As this chapter will elaborate, true danger in the war on terror is best defined by the more subtle categorization of areas that remain in shadow: neither governed nor controlled, with allegiance to no state in particular, these areas act like a no-man's-land ideal for terror groups. The Ferghana Valley is not dangerous because groups there will inevitably overthrow Kyrgyz, Tajik and Uzbek authority and form their own Islamic state, though this could be possible. It is most dangerous because for radical Islamists it is easy to get there, easy to stay and easy to leave as you please.

The Great Game and the Communist Revolution

By the mid-nineteenth century Central Asia was enveloped within the larger 'Great Game' diplomacy that pitted Great Britain against Russia. The British Empire was seeking to expand northward from India while Russia was seeking to expand southward from its Ural and Siberian territory. Central Asia was seen by both sides as a diplomatic and territorial symbol: some economic gain could be achieved no doubt but largely the acquisition was sought simply for the status it would represent for each empire respectively, thus the great game moniker. Ultimately the territory would fall under Russian imperial control. The Ferghana Valley in particular did not come under full control until 1876. Relatively ignorant and indifferent to the peoples, cultures and histories indigenous to the region, the Russian Empire made it a single administrative unit and designated it Turkestan.

Within a generation the Bolshevik revolution overran Russia which led to the emergence of the Union of Soviet Socialist Republics. Not surprisingly, this caused fundamental changes in the manner in which Central Asia was administered, but even then the volatility of the Ferghana Valley for Islamic revolutionary thought was readily apparent. Less than five years after the 1918 revolution it was necessary to dispatch Red Army brigades into the Ferghana Valley and surrounding areas to quell an uprising known as the Basmatchi movement. The Basmatchi were peasants who fought on horseback against the new plans of the command economy that called for forced collectivization and requisitioning.

Unlike other minor rebellions that were dispatched by the Red Army in the immediate years following the revolution, the Basmatchi movement was unique in that it was marked by nationalism and religious differentiation. It was clear from the very beginning that it was hatred for the non-Muslim occupiers of the Ferghana Valley that fueled the rebellion more than it was displeasure with collectivization. Loosely organized into small mobile groups, the Basmatchi were initially an even match for Red Army units because of their ability to freely cross borders and hide in the mountains of what is now present-day Iran and Afghanistan. Fascinatingly, it was their lack of unification under one command, caused by tribal and clan differences, which would ultimately signal their downfall as the Red Army presence increased. It is ironic to note that the very structure that spelled doom for the Basmatchi in the 1920s is exactly what makes their modern descendants so maddeningly difficult to locate and eliminate in the war on terror today.

'Uncle' Joe Stalin

By 1924 the Soviet leaders were aggressively beginning to redraw their new empire according to more strategic objectives. Stalin pushed especially hard for this in Central Asia, perhaps remembering some of the difficult particularities the Basmatchi had presented. Turkestan was thus divided into the present five states with borders decided arbitrarily and artificially. A perfect example of the divide-and-conquer rule, Stalin decided it would be advantageous to draw the borders in such a manner so that there would be large segments of non-titular nationalities inside of each new republic. Thus Kyrgyzstan enveloped a large segment of ethnic Uzbeks, Uzbekistan found itself home to a significant population of Tajiks, and Tajikistan took in thousands of Kyrgyz. On top of this, Stalin also decided to draw boundaries away from natural lines created by rivers and mountain ranges and made sure there was no ethnic or linguistic consistency across geographical areas.[1] Not surprisingly, the significance of these long-ago strategic decisions on present-day political issues is incredible. For example, the serious struggles today over water, gas and other resources are a legacy of this Stalinist strategy. The Kyrgyz and Tajiks control most of the sources for drinking water in the region. The Uzbeks and Turkmens have most of the gas.

It is not uncommon to arbitrarily draw borders as an imperial power; the mistakes of such imperial arrogance are readily seen today across the Middle East and Africa as well. But unlike those two regions, the Soviet Empire redrew the borders and then didn't make much of a concentrated effort enforcing adherence to them or pushing educational programs that effectively created an allegiance to these 'new republics'. As a result, the various tribes, clans and nationalities of

Central Asia freely traversed across these artificial 'national' boundaries. This was especially true in the Ferghana Valley where the tribal/clan tensions that had gone on for generations were basically unaltered by the political geography games being played in Moscow. Indeed, when one looks at a map of the Ferghana Valley states and the indigenous oblasts mixed between the three, it is hard not to think that the map was drawn with the intent to twist the three countries together as if tying a knot in a shoe.

Map 3.1 The Ferghana Valley and Native Oblasts[2]

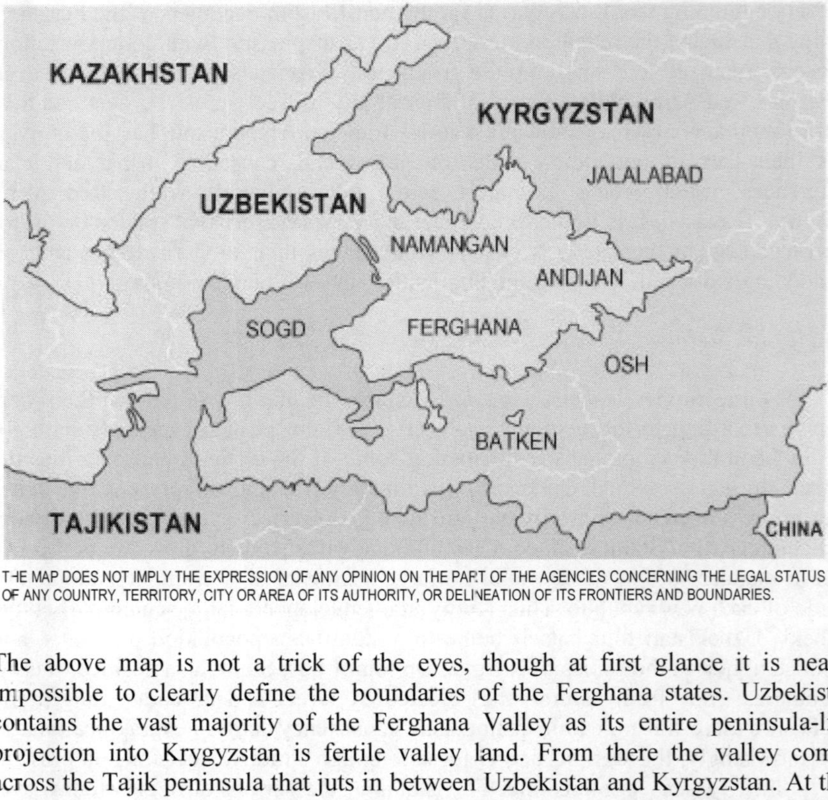

THE MAP DOES NOT IMPLY THE EXPRESSION OF ANY OPINION ON THE PART OF THE AGENCIES CONCERNING THE LEGAL STATUS OF ANY COUNTRY, TERRITORY, CITY OR AREA OF ITS AUTHORITY, OR DELINEATION OF ITS FRONTIERS AND BOUNDARIES.

The above map is not a trick of the eyes, though at first glance it is nearly impossible to clearly define the boundaries of the Ferghana states. Uzbekistan contains the vast majority of the Ferghana Valley as its entire peninsula-like projection into Krygyzstan is fertile valley land. From there the valley comes across the Tajik peninsula that juts in between Uzbekistan and Kyrgyzstan. At that point the valley then cuts sharply north back across Uzbekistan territory. Krygyzstan connects to the Ferghana Valley along the edges: its sovereign territory swoops under both the Uzbek and Tajik peninsulas and frames the outer eastern and southern edges of the Ferghana.

What all of this does in the end is create political authority chaos. To this day radical Islamist and criminal groups have taken advantage of this double-knot twist of countries in order to evade the law and maintain relative independence. It is this ability to freely migrate transnationally combined with the lack of jurisdictional oversight in each state domestically that makes the Ferghana Valley especially attractive to radical Islamist groups who find themselves adjacent to the valley from neighboring Iran, Afghanistan and Western China.

Demographics and Perestroikan Violence

The Ferghana Valley is small. It is about 180 miles long and 40 miles across at its widest point. Not surprisingly, surrounded as it is by relatively arid desert, the valley is packed tightly. The area represents just over four percent of Uzbekistan's overall territory but holds more than a quarter of its population, over five million people. The valley is also home to a third of all Tajiks and more than half of all Kyrgyz. In total the Ferghana Valley is home to more than ten million people. It is barely twice the size of Rhode Island.[3]

In the 1970s and 80s there was a demographic explosion across all of the Central Asian states but within the Ferghana Valley in particular. As a consequence, nearly 50 percent of the population in many of the Ferghana regions is under 30 today. This was especially troubling as the 1980s wound down and the Soviet economy ground to a relative halt: the youth population was hit the hardest by unemployment and the lack of economic mobility.[4] Add on to this the fact that the Ferghana Valley was home to more religious Muslims than any other part of Central Asia and by the early 1990s these factors – overpopulation dominated by youth, high unemployment and salient religious identification – combined to see radical Islamist thought make headway across the valley.

The most dominant form was easily Wahhabist, where Saudi Arabian propaganda and finances eagerly jumped into the ideological vacuum that emerged immediately after the fall of the Soviet Union. It was during this period that a mini-Islamic revival blossomed across the region. Dozens of mosques and madrassas, again largely financed with Saudi money, began popping up. Most are closed now and the how and the why will be discussed at length in the individual case study chapters. Most important for now is to recognize that it was the Ferghana Valley that played host to this blossoming. Since the early days of the Communist revolution if there was going to be radicalism along Islamic lines in Central Asia then that radicalism always found an eager audience in the Ferghana Valley.

Just as the Soviet Union was winding down, two incidents of serious bloodshed occurred in the Ferghana Valley. These two events, in 1989 in the Uzbekistan peninsula of the valley and in 1990 in the Osh region of Kyrgyzstan, showed how the ethnic/religious/tribal/clan volatility of the valley was susceptible to violence. Since these two events, however, there has been no bloodshed. Overall, this somewhat stunning absence of violence left the Western academic community relatively hopeful about the stability of the valley. Indeed, the foremost American expert on Central Asia, Martha Brill Olcott, noted that, 'Central Asia has suffered virtually every social ill: hyperinflation, rising unemployment, rising death rates, falling birth rates, deteriorating health care, government corruption and crumbling infrastructure which could be expected to increase social tension and so make inter-ethnic violence more likely, yet, Central Asia has recorded no large-scale ethnic-based disturbances since 1991'.[5] The present work argues that while this fact is indeed a point of hope for those studying ethnic violence, it is not politically wise nor academically sound to transfer this hope over to preventing Islamic radicalism. What has worked in the valley for preventing ethnic violence does not automatically equate to what is necessary to prevent a rise in Islamic fundamentalism. The details of these two ethnic violence incidents actually show analysts the potential for radical Islam to grow in Ferghana.

In 1989 there were over 100,000 Meshkhetian Turks living in Uzbekistan. Well more than half of those lived in Ferghana. The Meshkhetian Turks, an Islamic ethnic Turkish group historically from an area near the Georgian-Turkish border, were deported to Central Asia by Stalin during World War II.[6] For over forty years they lived peacefully within the valley and pursued a largely agrarian lifestyle relatively devoid of any political activism. This sudden violent clash in 1989 left, according to official statistics, 103 dead and over 1,000 wounded. Over 700 houses were either destroyed or severely damaged.[7]

In the end it remained somewhat of a mystery as to what exactly triggered this violent lashing out between Turks and Uzbeks who were closely related along both religious and linguistic lines, let alone the fact that the previous forty years had gone by with little tension. Ultimately, experts leaned on the catch-all political explanation of 'various underlying causes', which was basically to say the situation seemed so complicated and tangled it was doubtful anyone was going to be able to piece it together coherently and isolate a single causal factor. At most it opened up people's eyes to the idea that the Ferghana Valley was a tinderbox of potential violence that needed only an innocent spark to ignite.

It was feared that spark emerged one year later in 1990 when violence between Kyrgyz and Uzbeks in the Osh region left another 171 dead. 120 were Uzbek while 50 were Kyrgyz (the remaining one was Russian).[8] Just as had happened with the Turks and Uzbeks a year before, another 1,000 people were wounded with significant damage and destruction done to property. Just as before there was 'no discernable single event that could be isolated as the crucial moment that caused the outbreak of violence'.[9] Valery Tishkov, a former Russian Nationalities Minister and highly regarded scholar on ethnic violence, noted 'that the social environment and the communal culture can all contribute to precipitating the forces of violence'.[10]

While this work does not aim to call into question the judgment of such fine scholars as Olcott and Tishkov, it does wish to propose an alternative way for looking at the Ferghana violence in 1989 and 1990. Instead of being a confirmation of the difficulty in igniting violence, it should give cause for alarm because of the fact that violence emerged, such as it was, with 'no discernable reason or cause'. It seems at least somewhat viable to argue that given the right stimuli and preconditions, Ferghana's 'social environment and communal culture' is in fact ideal for housing Islamic radicalism movements. It is this potentiality that is most disturbing when considering the present global war against terror.

The Ferghana Valley faces numerous challenges: geographically, demographically, economically, and religiously. When trying to figure out, therefore, whether Ferghana should give us hope or concern it's important to keep in mind the difference between two dependent variables. This study isn't trying to ascertain what causes one ethnicity to act intolerantly towards another. When it comes to ethnic violence perhaps Olcott and Tishkov were correct and Ferghana remains a place of hope. This study is instead trying to ascertain where radical Islamism can not only emerge but flourish. In this context Ferghana sends off nothing but warning bells. The geographic, demographic and economic problems are the right preconditions. The emerging radical religious factor, a consequence of the repressive local governments and *exacerbated* by their relationships with the United States, could finally be the right stimulus. In the global war on terror you only need an environment that is both ruthlessly repressive and incompetently permissive. The governments of the Ferghana three have clearly done their foolish best to create such an environment.

Islam, Security and the Ferghana Three

In general terms the issues of Islam and security have been improperly analyzed by many scholars specializing in Central Asia. The tendency has been to focus on how the threat of Islam is exaggerated by the governments of Kyrgyzstan, Tajikistan and Uzbekistan, thus justifying their own repressive measures to ensure a longer, tighter grip on power. In addition Islamic radicalism within the region seems to always be judged according to whether or not a true threat can emerge that may actually topple or overthrow an existing government. 'True threat' is thus being defined in an extremely narrow and immediate way: only as a revolution-starter and with objectives to establish an Islamic theocratic state.[11]

By focusing only on this more classical form of threat (state overthrow) the Ferghana Valley does not necessarily emerge as a major geo-political hotspot. Consequently, most scholars consider the region's potential for radical Islamic identity to still be poorly-developed and too vague to mount a true challenge to the overpowering local regimes.[12] Even Ahmed Rashid, who wrote the highly praised *Jihad: The Rise of Political Islam in Central Asia*, had the majority of his detractors accuse him of underestimating locals' capacity to endure without rebelling.

This train of thought I find most confusing and naively dangerous: caution is ultimately the name of the game in combating terror and the last thing scholars should be doing is dismissing needed analysis with loosely-worded accusations of 'overestimating' threats or inexplicably relying on locals to endure abuse without consequence. Indeed, it is possible to wonder if the entire foreign policy of the United States when dealing with the Ferghana Valley is based on this very premise: as if short-term strategic advantages can be pursued to the detriment of long-term democratic development because it is assumed there won't be any harsh consequences capable of emerging beyond the local region. After all, the valley is supposedly home to people who *know how to endure punishment*. This seems to be shoddy diplomacy at best and incredibly misguided analysis upon which to hang hope.

This work attempts to correct this by broadening the analysis to include a more natural, subtle and likely danger: the world is better served when examination of radical Islam and the general political condition of the Ferghana three is concentrated around ascertaining whether the valley is not just a tinderbox for regional violence but whether it is also becoming a safe haven for larger transnational terrorist groups. These latter groups are not likely to focus on Kyrgyzstan, Tajikistan and Uzbekistan as exclusive targets. Thus, it is not so important whether Islamic groups in the valley aim to establish a revolutionary theocracy or seek to unite the entire region under the *Umma*, or Islamic community. In this new analytical light the initiatives of the Ferghana three in securing state power and the relations they maintain with western countries like the United States become much more important. In this light new Islamic threats emerge and should no longer be considered 'imagined' or summarily dismissed simply because their potential for violence is more gradual and less focused.[13] These characteristics are not signs of weakness or poor organization: they are purposeful strategies used by radical groups to remain in the shadows and out of the direct line of scrutiny.

A General Overview of Islamist Mobilization in Central Asia

The vast majority of scholars feel Islamist politics and potential insurgency across Central Asia, even within the volatile Ferghana Valley, is an exaggerated fear at best. Until 1999 it wasn't even considered a serious destabilizing factor as Islam's most important role up to that point was supposedly more involved with cultural revival than active governmental opposition.[14] Since 1999, however, significant and stunning changes have occurred across the valley regarding mobilization. In many ways the global war on terror has not acted as a stimulus to push radical Islamists deeper into inactivity but rather has elevated the Ferghana Valley as a desirable destination in which to develop a radical agenda.

Before 1999 policies that undermined the political salience of Islamic parties were actively pursued by the Ferghana three. Even Kyrgyzstan, which was often designated in the west throughout the 1990s as an 'island of democracy', did not permit parties to be formed exclusively on an Islamic foundation. Despite this heavy-handedness there were signs of burgeoning activity even then. It was during this so-called down period of Islamic activism that the Islamic Movement of Uzbekistan (IMU), long-regarded as the most openly radical and violent Islamic movement within Central Asia, was established. Where did the genesis of the IMU occur? It emerged out of the city of Namangan, one of the main Ferghana Valley cities in Uzbekistan.

Overall the Ferghana governments attempted to distinguish between those who 'belonged to mosques and schools subordinate to the semi-official religious administration' and those who did not.[15] The justification for such measures was at the time well-supported both within the former Soviet Union and internationally: with Chechnya looming large over everyone's heads, not just the Russians, all of the Central Asian governments rationalized the need for a strong hand and control over potentially dangerous issues like Islamic activism.

This general belief on the need to control and oversee Islamic groups seemed to be confirmed when six bombs exploded across Tashkent in February 1999. The International Crisis Group, a non-profit organization that acts as a watchdog over potential hotspots, best described the event in a report published two years later:

> Over roughly one and a half hours six bombs exploded, targeting symbolically important buildings, killing at least sixteen people and injuring over 100. One of the main explosions happened at the building of the Cabinet of Ministers as President Karimov's motorcade was about to arrive. In an astonishing breach of security, a car packed with explosives drove on to the premises despite strong security that always surrounded the President. Reputedly only a delay in arrival saved the President. The men who left the car and opened fire on building guards with automatic weapons escaped from the center of the most heavily protected area in the country. None of those who detonated any of the bombs were apprehended. This, combined with the precise timing, the high degree of coordination and security penetration, and the coincidence with other political developments led to immediate speculation that the attack was coordinated or carried out from within the security apparatus.
>
> The official response was to attribute the bombings to Islamist extremists. The day after the attack the Karimov government declared that it knew who was responsible and a series of arrests followed, some of which were carried out in other former Central Asian countries at the request of Uzbek authorities…The government of Uzbekistan has attributed the attacks to an international conspiracy of Islamist extremists, operating out of Afghanistan, Turkey, Tajikistan and the Ferghana Valley region of Uzbekistan itself. Many arrests for complicity were made in Tashkent and in the Ferghana Valley. The reputed conspirators abroad included most prominently Tohir Yuldosh, onetime leader of Adolat in Namangan (also in the Ferghana Valley) and now the political leader of the IMU, as well as Muhammad Solikh, the leader of the banned Uzbek opposition party Erk, who had been allowed to oppose Karimov in the presidential election of 1991 but had since been driven into exile. Though Solikh has denied any connection with the IMU, and the IMU itself has denied any responsibility for the bombings, Yuldosh has threatened that more bombings are in store if the Karimov regime does not step aside.[16]

The specific consequences these bombings had on Uzbekistan will be covered in the proceeding Uzbekistan chapter. The relevance for including it here deals with the symbolic turning point the Tashkent incident represented for Islamic activism across the Ferghana Valley. Up until that point in 1999 there was little evidence for Islamist groups to believe that the Ferghana states had any chink in their armor. Power seemed absolute and the security services were supposedly perfectly evolved from their KGB counterparts. In some cases they acted in ways that even surpassed the infamous precursor-institution. It was the Tashkent bombings that called out to all radical groups that despite repression and constant surveillance the state was not an impregnable fortress. It was after Tashkent that Islamic rhetoric, distributed across the valley either in the form of leaflets or audiocassettes, became much more virile in the hatred for the Ferghana regimes and more adamant in the need to fight and oppose these 'modern-day oppressors'. The Tashkent bombings not only gave hope to radical groups, it gave the first inkling of true fear to the home regimes.

A second incident occurred six months later in August 1999 that revealed not only how flammable the Ferghana area was but just how tenuous alliances were between the three Ferghana states themselves. This fact should not be underestimated: I have called the Ferghana Valley a 'no man's land for terror'. I argue this is one of the most favorable environments for radical Islamic terrorist groups to flourish because the area basically becomes ungovernable and pseudo-anarchic. Ironically, most of the heavy efforts by local regimes to reverse this trend work only to increase the advantages of the no man's land: as the governments of Kyrgyzstan, Tajikistan and Uzbekistan tightened their borders, increased security checkpoints and elevated alertness, their suspicion and mistrust of *each other* only exacerbated the ease with which security was undermined and overcome. This fact was vividly obvious in August 1999.

> On 6 August 1999 an armed group, which had entered the Batken District of southern Kyrgyzstan (also an oblast of the Ferghana Valley) through high mountain passes from Tajikistan, took several Kyrgyz hostages and released them a week later after a ransom of USD 50,000 had been paid. The Kyrgyz military proceeded with an effort to expel the armed group from Kyrgyzstan but on 22–23 August a new group of several dozen hostages were taken, including a Kyrgyz general and four Japanese geologists.
>
> The confrontation continued for two months, during which Kyrgyz troops appeared powerless to expel the insurgents from several villages. Uzbekistan offered military support and, without a go-ahead from either country, used its air force to bomb the territories of Kyrgyzstan and Tajikistan, provoking sharp protests. The insurgents finally withdrew after Japan reportedly paid a ransom of USD 6 million.
>
> Aside from the vague intention of bringing down the Uzbek government and the specific demand for ransom, the insurgents also insisted that the Kyrgyz government give them leave to pass freely through Kyrgyz territory to Uzbekistan, their actual primary target. Though the total number of fighters was reportedly near 1000, they could not realistically have expected that they would be able to take and hold any territory in Uzbekistan, which has… been fortifying border regions in recent years, especially in the Ferghana Valley.

As a result of these events, tensions grew sharply between the three states. Uzbekistan criticized Kyrgyzstan for military weakness and an inability to control its territory and accused Tajikistan of harboring Uzbekistan's enemies. Tajikistan protested the unsanctioned fly-over and bombing of its territory (which Uzbekistan first denied, then later admitted). Eventually, however, the incursion prompted security agreements between the states as they braced for new incursions.[17]

The incident raises an alarming number of red flags when it concerns the war on terror. The three-state incursion not only took place within the general political environment of religious repression, it occurred within the specific military context of increasing border fortification. The ensuing counter-measures taken to deal with this small 'invasion' involved border violations and the unilateral, arbitrary use of military force that violated the territorial sovereignty of two states. This was then followed up by accusations centering on conspiracy, weakness, cowardice and incompetence. The crisis in the end was 'averted' when the Japanese government intervened with the dubious strategy of actually paying a six million dollar ransom. While it worked in freeing its four citizens (which understandably was probably the only real priority for Japan), it was undoubtedly a bad precedent that left the Ferghana Valley ever more vulnerable in its wake.

These two 1999 incidents basically served as permission for the radicalization of Islamist groups. Considering how poorly the Ferghana states handled this mini-crisis, Islamist groups should have felt *compelled* to radicalize: with governments this incompetent and bickering, why not? When it comes to security breaches, the three Ferghana states are horrifically infantile in dealing with each other. The only thing they appear able to do is shift blame away from their own culpability and transfer it onto their neighbors. A simple rundown of the main conspiracy theories put forth at various times by all three Ferghana governments concerning Islamist opposition offers ample proof.

The main theory purports to account for Islamist opposition by designating it a ruse used by foreign criminals. Ultimately an ode to instrumental manipulation, this international conspiracy is operated and financed by those foreign cartels most interested in expanding and deepening their transit routes and bases of operation for narcotics-trafficking.[18] With this version it is possible to discount Islamic opposition altogether because the most radical groups are not driven by sincere belief or faith in ideology but instead use Islam as a pretense in which to attract and then co-opt Ferghana youth into the criminal world. It is for this reason, pre-9/11, Ferghana states often simply refused to allow the use of the term 'Islamist' when considering such reports. This relatively arrogant dismissiveness was not exclusive to the governments sharing Ferghana. A major member of Interpol during a conference in Tashkent declared flatly that the insurgents' Islamist ideology was merely a smokescreen for drug-trafficking.[19]

A second very popular theory moves Islamist radicalism out of the criminal world and plants it firmly on the shoulders of foreign governments politically conspiring to undermine and destabilize the region. The usual culprit in this is

Pakistan, which is accused of actually *wishing* to see a spread of Taliban-like regimes across the region since it would only increase its leverage and relevance on the world stage, especially vis-à-vis the United States, as an ally. The other guilty parties in this line of reasoning are Saudi Arabia and the Gulf states. They would like to see their spheres of influence expand outward from the Middle East. At least this second conspiracy theory has some train of logic: Saudi Arabia has actively financed Wahhabist propaganda and encouraged the building and opening of madrassas within the valley. But this does not, however, amount to a concerted covert effort to overthrow the region and turn it into a Saudi Arabian annex.

The third and final conspiracy theory falls back on an old whipping boy: Russia. Ironic considering Russia's decade-long problem with radical Islamists in Chechnya, this idea accuses Russia of encouraging a radicalization of Islam and turning an indifferent eye to foreign nationals looking to cross borders into Ferghana. What is the ultimate objective of this plan? To destabilize the region enough so that the five states of Central Asia would be clamoring for greater assistance and a tighter relationship with Russia. In this manner Russia would be able to reconstitute its regional dominance and regain a semblance of its old superpower status.

While Russia is of course not exactly the most innocent of states when it comes to angling for political leverage across the former Soviet Union, this theory is the least logical. The fact of the matter is that most military circles within Russia are convinced that radical Islamists filter through Central Asia *in order to get to Chechnya and kill Russian troops*. Looking to destabilize the Central Asian region with Islamist terrorists who would come to Ferghana and then remain there is just not fathomable to Russians. They are convinced the ultimate goal would be to continue across the Caspian and on into Chechnya. Destabilizing the valley in this way, therefore, would only further exacerbate Russia's own terrorist problem.

While the reporting on the 1999 incidents by the International Crisis Group was top-notch it should be noted that it also in the end missed a more dangerous and relevant conclusion, although this is at least explainable by the fact that the report came out before 9/11. Its ultimate conclusion regarded the incidents as 'not threatening to bring down governments in the near term. The more significant threat they represented was the possibility of a more widely popular uprising that would both feed on and exacerbate serious tensions within the countries'.[20] 9/11 brought a new perspective to these conclusions, revealing a new threat: these incidents decisively symbolized the relative ease with which a group could base operations out of the Ferghana Valley. Not only were there cultural, religious and linguistic compatibilities within the valley, there was also a strong historical tendency to at least be sympathetic to radical Islamists. On top of it all the utter idiocy with which the local governments reacted only served to prove that regional defenses for the valley were easily surmounted. Finally, the continued stubbornness of Ferghana states to entertain conspiracy theories and their penchant for looking exclusively abroad for root causes only made the region more attractive because such adolescent games only distracted local forces from real security. Indeed, this work shows that Islamist groups learned these lessons well in 1999.

The political geography and climate in the Ferghana Valley make the area an ideal locale for developing future transnational terror threats even if groups in the valley itself remain relatively weak in trying to overthrow local regimes. One other factor must also be emphasized to more convincingly support the relevance of the valley: the harsh treatment by the Ferghana governments creates an atmosphere of complicity with if not sympathy for radical groups who may seek safe harbor. This simple fact has proven to be an incredibly difficult obstacle to overcome, both militarily and politically for the United States, in Afghanistan and Iraq. Unfortunately the same environment is clearly developing within the Ferghana Valley. If it was ever necessary to conduct operations there US forces would experience the same barriers to success.

As with the flaw in analyzing radical Islamist groups strictly according to their ability to produce a successful revolution, it is a mistake to analyze the impact of harsh government measures against the people exclusively along lines of revolt and civil unrest. Since this has in fact been the focus of the vast majority of studies that look into the Ferghana Valley and Islamic mobilization, the inevitable conclusions vary from acquiescence to apathy: we do not find massive civil protests or industry-wide strikes or meeting halls filled to the brink every Friday night with angry citizens. Focusing exclusively on this, however, is a mistake simply because *it is unnecessary for a popular revolt to happen in the Ferghana Valley in order to make the Ferghana Valley the next great center of Islamic transnational terrorist development*.

The diplomatic community has forgotten that apathy can also be permissiveness. The simple willingness to not comply with government operations can be a huge advantage in favor of terrorist groups. The treatment of the people by the Ferghana three has been so harsh and arbitrary, especially in the realm of religion, that there has been a severe polarizing effect within valley communities. Has it been enough to make regular everyday people want to risk their lives in what would most likely be suicide against government security agencies? No, it has not but it has been enough to make the general population massively in opposition to the state and consequently against any allies who are seen as tacit supporters of the state. This level of hatred and contempt has unfortunately already been breached.

The three Ferghana states have not only been ruthless in arbitrarily persecuting the Islamic faith, they have also been extremely aggressive in trying to co-opt Islam so that they are better able to monitor and subvert worship itself. After all, all three governments openly proclaim to be Islamic. Thus, there is a dual negative impact on society by way of governmental repression: not only is the general population shunted toward sympathy for if not outright participation with radicals, those who would never consider themselves radical or imagine themselves supporting radical causes ultimately move away from mainstream Islamic institutions because of the societal suspicion that all such institutions are compromised. Many across the Ferghana Valley believe that there is no reason to fear only if you openly cooperate with the government. The problem is that no one is going to come and worship if it is understood this mosque or this imam is a cooperator with repressors. It is this atmosphere that makes neutrals suddenly stand on the side of radicals. It is not that they necessarily wish to be there or find themselves to be true believers in the radicals' cause. It is simply the fact that in this environment it is the radicals who appear to have a version of Islamic faith that is more legitimate and less politically corrupt. It is this final aspect, in combination with political geography and climate, which makes the Ferghana Valley such an ideal chemistry set waiting for an Islamic explosion. That chemistry set, though yet to explode, has been increasingly shaken by 9/11 and its political aftermath.

Perspectives on 9/11: Mixed Signals and Missed Opportunities

Before the ashes and dust even settled on the wounded ground of New York City on September 11, 2001, it was clear to the Ferghana three that the collapse of the Twin Towers was an incredible opportunity. All three ended up supporters in America's war against terror. That support, however, varied in intensity over the next few years largely because of internal considerations and what each stood to gain from their respective cooperation. Uzbekistan was clearly the most enthusiastic and saw the opportunity to tie its own radical Islamic IMU problem to Al-Qaeda even faster than Vladimir Putin did in trying to tie the Chechens to Osama bin Laden. It was obvious that Uzbekistan hoped to bag two birds with one stone: not only would it be an ingenious stroke of good luck to get the United States collaterally inflicting damage on the IMU but Uzbek cooperation would likely garner much needed economic and political concessions that would do wonders in shoring up President Karimov's regime.

Tajikistan felt motivated to offer support immediately in the aftermath but also was quick to couch this support in much more temperate muted language. There were no doubt serious concerns about how a war against radical Islamists, perceived locally as being a purely American-led endeavor, would play at home where the Tajik government was still not far removed from ending its own nasty civil war (in 1997) that carried deep religious undertones of tension and was now operating in a loosely-bound secular-Islamic coalition. This coalition was by no means on solid ground in the fall of 2001 (just how insecure and real this coalition was will be discussed in detail in the Tajik chapter) and it was clearly feared that Tajik cooperation could ultimately backfire on the local effort to maintain peace and stability.

Kyrgyzstan for its own reasons was also supportive but wary of the long-term consequences, especially the potential influx of refugees flooding into the region once the attack on Afghanistan began. This was always somewhat curious considering Kygryzstan did not actually share a border with Afghanistan. Interestingly, Kyrgyzstan already had a well-founded commitment to fighting terror, having its capital selected by the Shanghai Cooperation Organization (a loose regional organization comprised of Russia, China, Kazakhstan, Uzbekistan, Tajikistan and Kyrgyzstan) for headquartering a joint anti-terrorist center.[21] This priority ultimately won out over cautiousness as Kyrgyzstan territory became a major instrument in American operations against the Taliban in Afghanistan.

Kyrgyzstan

While Kyrgyzstan held no love for radical Islamic groups President Akaev's regime was generally reluctant at first to show anything more than the obligatory condemnation of the 9/11 attacks. This was mainly on account of the intense yet largely unfounded fear within the capital that the country would somehow get physically embroiled in the ensuing Afghan crisis. Always having been militarily weak and extremely dependent on the larger neighboring powers of Russia and China, Kyrgyzstan did not see as many immediate advantages as Uzbekistan did in carving out a major role in the global war against terror.

As international public consensus continued to build in the aftermath of 9/11 that Osama bin Laden was the main financier and organizer, Kyrgyzstan finally found common interests with the US in actions against Afghanistan: several IMU incursions into the sovereign territory of Kyrgyzstan over the years had been rumored to be financed by Osama bin Laden. If the Americans were interested in using Kyrgyz air space or launching off of Kyrgyz territory in order to destroy bin Laden and his base of operations in Afghanistan, this would only benefit Kygryzstan in the long-run.

This motivation did ultimately carry the day but the dangers that could emerge for Kyrgyzstan should Afghanistan implode into a smoldering pile of anarchy were never far from the minds of Bishkek authorities. Colonel Oleg Chechel, Chief of International Military Cooperation at the time, explained succinctly the subtleties of Kyrgyz cooperation.

> Future developments will depend on where and what kind of strikes will be conducted. If we talk about military operations against vast territories of Afghanistan then we could see an inflow of hundreds of thousands of Afghan refugees into our region. Pakistan has blocked its border with Afghanistan, Iran will not accept any sizeable group of refugees so the only direction for Afghans is north – Tajikistan, Kyrgyzstan and Uzbekistan. Besides problems with accommodating refugees we might face members of terrorist groups and intelligence organizations of various countries infiltrating Kyrgyzstan.[22]

Tajikistan

Just like Kyrgyzstan, Tajikistan was not able to act completely independent of Russia in terms of military operations since the fall of the Soviet Union. In 2001 Tajikistan remained the only Central Asian country in which Russian ground forces and border guards were still stationed, with some sources citing as many as 20,000 troops. Much of the Tajik border patrol duty had always fallen upon Russian-led or Russian-controlled military units. Remarkably, at the time of the 9/11 attack, only 70 kilometers of the 1,300 kilometer long Tajik-Afghan border was actually patrolled by Tajikistan's own army.[23]

President Rakhmanov had long been one of the most strident supporters of fighting terrorism within the former states of the Soviet Union. But this passion had to be tempered when it concerned open support for American operations in the area. Not only would Tajikistan most likely be more legitimately affected by a refugee problem should things go badly in Afghanistan, it simply had to take into account the Russian military presence on the ground and the foreign policy wishes of Moscow, which was not overly excited by an increased American presence in Central Asia.

On top of all of this was the loosely-based governing coalition formed after the Tajik civil war between secular elites and the Islamic Renaissance Party of Tajikistan. One of the few so-called positives to emerge from the civil war, Tajik Islamists tenuously agreed to remain on a path of open political channels and operating exclusively within the existing system. This supposedly helped end the civil war because finally real compromise could be achieved. The system supported by the Taliban, therefore, was justifiably anathema to the secular elites who worried of ideological spillover. Thus of the three Ferghana states it was clearly Tajikistan authorities who had to draw the most precise balance, not only between Moscow and Washington but between local secular and Islamist forces.

Uzbekistan

More so than the other two, Uzbekistan was the one Ferghana power that seemed intent from the start in capitalizing on 9/11. Easily the most powerful and militarily independent of the Ferghana three, President Karimov did not necessarily have to consult with Moscow or Washington for advice. Thus it was clear for Karimov that cooperation with America's anti-terror war could be an effective tool for distracting the rising international criticism that had emerged against his government since the Tashkent bombings in 1999.[24]

Fascinatingly, post-9/11 American operations created within Uzbekistan a microcosm of the larger problems that would later afflict the Ferghana region as a whole. Critics of Karimov were always worried that 'out-of-sight, out-of-mind' could become a bitter mantra of the international community. In fact many critics within Uzbekistan felt the United States could ultimately undermine its own long-term anti-terror objectives if it were to ally itself too closely to Karimov. One Uzbek official, wishing to remain anonymous, put it aptly:

> The US government will fight the Islamic terrorists and our government will get full support from the West to fight against *those our government declares terrorists*. Since the West has little understanding or interest in distinguishing between devoted Muslims and extremists or terrorists all opponents of the government will be easily jailed.[25]

This concern, at the time strictly Uzbek in origination, would in fact come to epitomize the problem for all of Ferghana vis-à-vis the global war on terror. Many inside of Kyrgyzstan, Tajikistan and Uzbekistan rightly saw through their own governments' slyness and knew that 9/11 was ultimately just a piece in a larger local Machiavellian puzzle. The true fear for them was that the United States would not see the ploy or, even worse, not care. Indeed, nearly five years into the war on terror, it seems that fear was well-placed.

NATO and the Jets

By 2005 the war on terror still raged on and the cooperation from the Ferghana three was well-established and considered a crucial part in successful operations. It was hoped by scholars that 2005 would be a seminal period as all three Ferghana states were scheduled to have either parliamentary or presidential elections. So what sort of democratic progress was achieved as a consequence of this 9/11 cooperation and engagement by the US? In short, there was no progress and a quick analysis of media freedom across the three states illustrates it compellingly.

Kyrgyzstan's capital, Bishkek, lost its popular and only independent television station, Pyramida, when station management sold the majority of shares to a government affiliate late in summer 2004. As a consequence, Kyrgyzstan conveniently had no independent media sources in the run-up to parliamentary and presidential elections.[26]

With Tajikistan's parliamentary elections only few months away in early 2005 the government began crudely persecuting most of the main print media in the capital of Dushanbe. Despite the secular-Islamic governing coalition being constantly lauded as a model for the rest of Central Asia, it did not stop the printing press of the Islamic Renaissance Party from being illegally shut down over the summer for supposed tax violations. This effectively stopped the distribution of the two biggest opposition newspapers in Tajikistan, Ruzi Nav and Nerui Sukhan. This came after the Ruzi Nav editor was attacked and beaten outside of his home.[27]

Uzbekistan's parliamentary elections, set for the end of 2004, prompted an overall strategy of suppressing and pressuring the media. The media-support organization Internews Uzbekistan was shut down by court order for six months. Five partner stations were all stripped of their licenses. Internews Uzbekistan's director, Khalida Anarbaeva, decried the attacks as blunt attempts to undermine a free and independent media in Uzbekistan before election campaigns were begun.[28]

More subtle and damning to this blatant government censorship were the accompanying changes in electoral tax legislation that ultimately codified and institutionalized interference. Tajikistan had not even passed a new election code and deftly manipulated its tax regulations in order to pick and choose tax 'victims'. Kyrgyzstan's election law was cleverly restrictive, forbidding campaigning 'through foreign media'. This ultimately carried repercussions that saw Russian-based newspapers and such venerable news agencies as the BBC and Radio Liberty limited if not denied access to covering Kyrgyz elections. Uzbekistan developed a 'working group' for parliamentary journalists that served as a partial control mechanism during elections.[29]

Kyrgyzstan's broadcast licensing ground to a halt a full year before elections while Uzbekistan simply refused to register any new media outlets for months at a time. Tajikistan also did not register any independent media stations in its capital and developed a tendency to simply let formal applications from foreign media twist on the vine, often for as long as three years. For those media outlets actually 'lucky' enough to have maneuvered through this labyrinthine mess of purposeful subterfuge they were greeted by an environment in which judicial branches looked extremely favorably upon libel, defamation and 'protection of dignity and honor' cases when filed on behalf of the government. Kyrgyzstan and Uzbekistan were the first to make wide-spread use of such suits to muzzle legally-registered media sources and by 2004 Tajikistan had caught up to its Ferghana brethren.[30]

This was but the tip of the political environment iceberg in the Ferghana Valley. The proceeding chapters reveal the massive part that remained relatively hidden under an ocean of international ignorance and indifference. Still it is important to note briefly here just how the international community interacted with and observed the Ferghana states during post-9/11 operations. In late 2004 NATO Secretary-General Jaap de Hoop Scheffer visited Central Asia and gave praise lauding each state for its cooperation in the war on terror. Scheffer was also adamant about NATO's hope for continued close involvement between the security organization and each of the Ferghana states. Kyrgyzstan wasted no time at all during Scheffer's visit, specifically requesting that NATO help re-equip its army and begin supplying modern weapons to the republic. Scheffer lauded Tajikistan for signing an historic transit agreement with NATO that would support ISAF forces deployed in Afghanistan. While tactical strategy seemed to dominate Scheffer's visit as NATO considered the possibility of opening training centers in the Central Asian states, the most telling moment of the Secretary-General's entire trip was a simple single line given in response to how committed NATO, and consequently the West, was in maintaining an alliance with the most powerful Ferghana state despite reports of its uneven democratic development:

> The human rights situation in Uzbekistan will not anyhow hinder cooperation between NATO and Uzbekistan.[31]

Leave it to a military general to state the obvious most succinctly. While American diplomats obfuscated and Western scholars pondered imagined dilemmas, Secretary General Scheffer made it crystal clear to the Ferghana three: ultimately all that really mattered to the West was that you continued to provide operational assistance in the global war against terror. As the ensuing chapters will testify, Kyrgyzstan, Tajikistan and Uzbekistan were listening quite intently.

Conclusion: Bubble, Bubble, Toil and Trouble

As academics we have focused too exclusively on the Ferghana Valley itself, thinking any Islamic fire to emerge from there will have to first burn bright within the valley before being able to spread out onto the global stage. This is simply wrong because of the structure and nature of contemporary Islamic terrorist groups. They do not necessarily have to wage war with local regimes before seeking larger international targets. In fact, they are better off if they do not have to. Indifference and isolation in the home region is ideal for Islamic radicalism. This chapter has attested to the fact that in spite of oppressive regimes the Ferghana Valley has remained such a place and if anything has become even more optimal since 9/11.

Central Asian policy took advantage of the Twin Towers collapse and ultimately pushed mainstream Ferghana society further to the edges of Islamic radicalism. Economic conditions have not improved. Moderate Islam has been co-opted to the point of illegitimacy. The futile quest for justice hangs over society like some unreachable Holy Grail. To date only radical Islamists have come close to offering remedies to these problems while the indigenous governments and the international community have looked particularly indifferent and subsequently cruel.

The conclusion of the International Crisis Group's report on Islamic mobilization, written *before* 9/11, is timely and pathetic in its prophetic frustration:

> The greatest risk represented by [radical Islamists] is that they can become identified with wider popular aspirations for better economic conditions, a more Islamic society or a quest for justice…Foreign governments and international organizations must support unwaveringly and unequivocally the position that regional security can only be assured if religious freedom is guaranteed and the legitimate activities of groups and individuals are not suppressed…Any support to the region's governments with the goal of enhancing security will be futile if the factors leading to polarization of the population against the governments are not reduced or eliminated. *Donors should be very careful to ensure that their activities do not strengthen or legitimate…Islamic groups.*[32] (Italics mine).

In a nutshell, that is exactly what has been happening since 9/11. The supposedly overwhelming need to ensure immediate security has compromised more rational and long-term visions for engaging the Islamic world in ways that would reduce and ultimately eliminate the viability of terrorism. Local regimes, foreign governments and international organizations all worked together, sometimes by design, sometimes unknowingly, to create a cauldron of potential trouble in the Ferghana Valley. This cauldron mixed together some pretty potent chemicals:

- The lack of religious freedom
- The increase of complicit support from non-radicals
- The pressure of intensifying poverty
- The sense of an international betrayal of justice.

Domestic and international activities did indeed intensify the suppression of Islam. As an end result, however, it only legitimized the most radical purveyors of the faith. And while the cauldron of Ferghana is not yet boiling that should offer the global community little relief: if and when it finally does bubble over it will most likely not burn in the valley. The cauldron of Ferghana will most likely begin to burn in the West.

1 Bob Kaiser and Lois Raimondo, 'The Volatile Ferghana Valley', *The Washington Post*, 15 July 2002.

2 United Nations Development Program, 'Environment and Security: Transforming Risks into Cooperation', http://www.envsec.org/centasia/maps.php, accessed 13 November 2005.

3 Bob Kaiser and Lois Raimondo, 'The Volatile Ferghana Valley', *The Washington Post*, 15 July 2002.

4 Eisenhower Institute's Center for Political and Strategic Studies, 'An Overview of the Ferghana Valley', Volume 1, Number 1, January 1996.

5 Eisenhower Institute's Center for Political and Strategic Studies, 'An Overview of the Ferghana Valley', Volume 1, Number 1, January 1996.

6 Eisenhower Institute's Center for Political and Strategic Studies, 'An Overview of the Ferghana Valley', Volume 1, Number 1, January 1996.

7 Eisenhower Institute's Center for Political and Strategic Studies, 'An Overview of the Ferghana Valley', Volume 1, Number 1, January 1996.

8 Eisenhower Institute's Center for Political and Strategic Studies, 'An Overview of the Ferghana Valley', Volume 1, Number 1, January 1996.

9 Eisenhower Institute's Center for Political and Strategic Studies, 'An Overview of the Ferghana Valley', Volume 1, Number 1, January 1996.

10 Eisenhower Institute's Center for Political and Strategic Studies, 'An Overview of the Ferghana Valley', Volume 1, Number 1, January 1996.

11 Dr. Shahram Akbarzadeh, Central Asia-Caucasus Analyst, 'Is Islam a Threat to Stability in Central Asia?', Johns Hopkins University, http://www.cacianalyst.org/view_article.php?articleid=188, accessed 12 July 2004.

12 Dr. Shahram Akbarzadeh, Central Asia-Caucasus Analyst, 'Is Islam a Threat to Stability in Central Asia?', Johns Hopkins University, http://www.cacianalyst.org/view_article.php?articleid=188, accessed 12 July 2004.

13 Dr. Shahram Akbarzadeh, Central Asia-Caucasus Analyst, 'Is Islam a Threat to Stability in Central Asia?', Johns Hopkins University, http://www.cacianalyst.org/view_article.php?articleid=188, accessed 12 July 2004.

14 International Crisis Group, 'Central Asia: Islamist Mobilization and Regional Security', ICG Asia Report No. 14, 1 March 2001.

15 International Crisis Group, 'Central Asia: Islamist Mobilization and Regional Security', ICG Asia Report No. 14, 1 March 2001.

16 International Crisis Group, 'Central Asia: Islamist Mobilization and Regional Security', ICG Asia Report No. 14, 1 March 2001.

17 International Crisis Group, 'Central Asia: Islamist Mobilization and Regional Security', ICG Asia Report No. 14, 1 March 2001.

18 International Crisis Group, 'Central Asia: Islamist Mobilization and Regional Security', ICG Asia Report No. 14, 1 March 2001.

19 International Crisis Group, 'Central Asia: Islamist Mobilization and Regional Security', ICG Asia Report No. 14, 1 March 2001.

20 International Crisis Group, 'Central Asia: Islamist Mobilization and Regional Security', ICG Asia Report No. 14, 1 March 2001.

21 International Crisis Group, 'Central Asian Perspectives on 11 September and the Afghan Crisis', Osh/Brussels, 28 September 2001.

22 International Crisis Group, 'Central Asian Perspectives on 11 September and the Afghan Crisis', Osh/Brussels, 28 September 2001.

23 International Crisis Group, 'Central Asian Perspectives on 11 September and the Afghan Crisis', Osh/Brussels, 28 September 2001.

24 International Crisis Group, 'Central Asian Perspectives on 11 September and the Afghan Crisis', Osh/Brussels, 28 September 2001.

25 International Crisis Group, 'Central Asian Perspectives on 11 September and the Afghan Crisis', Osh/Brussels, 28 September 2001.

26 Central Asia-Caucasus Analyst, 'Media and Elections in Central Asia', Johns Hopkins University, http://www.cacianalyst.org/view_article.php?articleid=2707, accessed 24 September 2004.

27 Central Asia-Caucasus Analyst, 'Media and Elections in Central Asia', Johns Hopkins University, http://www.cacianalyst.org/view_article.php?articleid=2707, accessed 24 September 2004.

28 Central Asia-Caucasus Analyst, 'Media and Elections in Central Asia', Johns Hopkins University, http://www.cacianalyst.org/view_article.php?articleid=2707, accessed 24 September 2004.

29 Central Asia-Caucasus Analyst, 'Media and Elections in Central Asia', Johns Hopkins University, http://www.cacianalyst.org/view_article.php?articleid=2707, accessed 24 September 2004.

30 Central Asia-Caucasus Analyst, 'Media and Elections in Central Asia', Johns Hopkins University, http://www.cacianalyst.org/view_article.php?articleid=2707, accessed 24 September 2004.

31 Central Asia-Caucasus Analyst, 'Media and Elections in Central Asia', Johns Hopkins University, http://www.cacianalyst.org/view_article.php?articleid=2707, accessed 24 September 2004.

32 International Crisis Group, 'Central Asia: Islamist Mobilization and Regional Security', ICG Asia Report No. 14, 1 March 2001.

Chapter 4
Kyrgyzstan: Misreading a Revolution

The Republic of Kyrgyzstan represents a roller-coaster of international assessment since its independence in 1991. Initially, it was considered a true 'island of democracy' surrounded by other more repressive Central Asian governments. Over time that designation seemed ill-fitting as the government of President Askar Akaev was seen to shift during the mid-1990s to a regime more consistent with the area: one exemplified by a political system thoroughly corrupt, security forces increasingly more ruthless and brutal, with popular protest ever more isolated and ignored that in turn created radicalized opposition. This chapter ultimately reveals that the 'island of democracy' tag was always a misnomer and that more recent events, including a so-called democratic revolution that earned support from the United States, only continued disturbing trends within Kyrgyzstan that can only do harm in the bigger picture of eliminating the causes of radical Islamism.

The 'Island of Democracy'

Immediately after independence, coinciding with the fall of the Soviet Union, Akaev pushed hard to create ties with neighboring countries and talked of staunch commitment to liberal, free-market reforms. It was this stance more than any other that made him such a favorite of the international donor community inside of Central Asia. But this, if anything, was only proof of the international community's tendency to conflate and confuse economic liberalization with democratic reform. The 'island of democracy' title was bestowed initially upon Akaev not so much for there ever being a commitment to true democratic development but instead for supposedly launching quick liberalization programs that opened Kyrgyzstan up to foreign corporations. This continues to be a common mistake in how the western diplomatic community deals with the transitioning world. This is no small matter as economic hardship combined with massive levels of corruption always emerges from this sort of miscalculation.

Ironically, there was trouble for the island of democracy even before Kyrgyzstan became an independent state. In 1990 Kyrgyz intelligentsia, riding the wave of peaceful change that had been powering across the entire Soviet Union, coalesced into the Democratic Movement of Kyrgyzstan (DMK).[1] It was only two years after independence that such executive-legislative branch harmony changed into political discord. The new Kyrgyz parliament, dominated by the DMK, grew fiercely resistant to what it considered rampant corruption and mismanagement within the Akaev government. Akaev responded in turn by proposing a public referendum for approval of his presidency. This was a move actually perfected in Russia by Boris Yeltsin because in this early era of post-Soviet transition executive heads of state held an incredulous amount of influence. No one would be able to compete with Akaev simply because there was no figure with as large a stature or with as much unfettered media access as the executive leader.

In January 1994 the presidential referendum results were announced as having garnered 96 percent in favor. Though the procedure for holding the referendum was largely seen as falsified the ultimate consequence was dissolving the parliament. A new bicameral parliament, the Jorgorku Kenesh, was created and not surprisingly made totally subservient to Akaev. The DMK's leader, Topchubek Turgunaliev, was imprisoned on charges of fraud.[2] So within two years of new freedom, Central Asia's island of democracy had already effectively neutered the principle of checks and balances. The international community was largely silent, a trend that would disturbingly continue as Akaev would enlarge the extent of executive power in a series of initiatives: the referendum of 1994 was followed by referenda in 1996, 1998 and again in 2002, all intensifying the grip of presidential power. They ostensibly ensured that the personal careers of almost every government employee were beholden to Akaev for appointments.[3]

In this aftermath opposition leaders adopted a more subtle strategy: they began channeling their criticism through private media outlets and non-governmental civil society organizations.[4] Before this was able to establish a strong foothold in Kyrgyz society, however, the Akaev government launched its own counter-measures: it successfully consolidated most major printing presses into state ownership and closed three popular newspapers entirely.[5] Government officials also began at this time the habit of filing suits against local newspapers, thereby exposing the media to charges of slander, libel and defamation of character. Most suits were brought strictly according to the wishes of President Akaev.

Incentives and Penalties

Akaev not only wielded a big stick when it came to ensuring his own grip on power. His government from the very beginning applied a varied strategy of both stick and carrot to achieve its goals. Indeed, co-optation was probably the most favored strategy of the Akaev government for it widened the circle of culpability, making Akaev less vulnerable to individual criticism. For incentives governmental officials often granted tax relief, financial concessions and state contracts to the businesses of prominent figures in return for allegiance to the regime. Akaev, for example, issued a decree in 1994 that offered monthly salaries to aksakals, local elders usually instrumental in the development of grassroots activism within Kyrgyz society. He also made key appointments to the presidential cabinet of some outspoken civil society leaders.[6] Opposition figures that were regarded as moderate were dealt with even softer velvet gloves by granting them free vacations, chauffeured government cars and cheap property in prestigious Bishkek regions.[7]

It was clear that cabinet appointments carried no real power or decision-making capability: they were mere buy-offs meant to purchase the opposition and thus quell dissent. It was a hallmark of the Akaev regime and should give readers pause when considering just how 'liberating' the initial Kyrgyz economic liberalization programs were at the start of the 1990s. Ironically, it was these reforms, and not any democratic program, that earned Kyrgyzstan the nickname as an island of democracy. In reality, even these 'positive steps' in Kyrgyz development only served as instruments for undermining civil society and undoubtedly crushed any real development of a middle class. It was just such corrupt liberalization that created many of the oligarchs that now plague the former Soviet Union.

Penalties that became common during Kyrgyzstan's 'island' years included a vast array of bureaucratic weapons. Aside from the more dramatic imprisonment of such figures as Feliks Kulov and Azimbek Beknazarov there was also continuous harassment, including assault and forced exile. This utilization of the judicial system to do the executive branch's bidding was quite common. Additionally, the tax inspectorate and other financial agencies would consistently attack opposition leaders and members of the mass media with aggressive audits.[8] Zamira Sydikova, editor of the weekly Respublica, for example, was found guilty of libel and slander three times, in 1995, 1997, and 2000.[9]

Though most of western academia declared Kyrgyzstan an island of democracy well into the 1990s, the relative abandonment of such praise today is still usually accompanied by a disclaimer attesting to the fact that Akaev gradually began undermining his own reforms. This is simply not true and it remains something of a mystery as to why Western scholars have been slow to acknowledge this fact. Kyrgyz political development immediately after the fall of the Soviet Union was marked by Akaev pursuing policies and engaging in strategies that created severe consequences wholly detrimental to democracy: the judiciary came under the control of the presidency and was frequently used to silence opponents; the media, far from being one of the freest in the region, was under constant pressure and frequently subjected to invasive tax audits and corporate takeovers initiated by the government; civil society organizations were continuously harassed, often by police and security apparatuses that were increasingly harsh and infamous for their corruption; the economy became dominated by monopolies, controlled by oligarchs and usually tied directly to the presidential family and its allies.[10]

Clearly, the Kyrgyz island of democracy was built on a foundation of quicksand. Closer to a feudal system than a democratic one, Akaev's government always hovered near impending crisis because of the manner in which its executive branch dealt with both governance and representation. There was never an initial positive period that truly represented an emerging democratic system.

Governance and Representation

A crisis of governance can occur when there is a lack of decision-making capacity in the highest political posts.[11] This was the case in Kyrgyzstan as Akaev purposefully rigged the system to be accountable to him and only him. While this may have made sense for Akaev it had devastating effects on the system as a whole as policy paralysis very quickly overran almost all agencies. In a system characterized by such utter powerlessness and irresponsibility, in time no one looked to make decisions for which they would be held accountable. Instead they looked to escape accountability. Obviously this did not occur in a vacuum, with the most immediate consequence in Kyrgyzstan being the total lack of trust in society for its national government.

This was only exacerbated by the widespread practice of buying and selling posts.[12] The easiest manner in which to ensure allegiance was to create a system that had very few directly-elected positions. With the power of appointment the president had the potential to control any system with impunity. Since Akaev didn't want any major decision-making power to exist outside of his direct control the only reason to gain a political appointment was in order to accumulate personal riches and curry favor. Abuse was so widespread that a common societal complaint became no one in government had neither time nor inclination to solve actual problems: they were all too busy trying to 'earn' their personal fortunes.

Ultimately this was not solely about the isolation of power from the people. Unfortunately for regular Kyrgyz, they were *never* a part of the political equation. Rather this system cultivated by Akaev was about limiting the access of various elite groups to power. By keeping power and wealth limited to an incredibly small circle of favorites, Akaev was always playing with political fire. As one southern Kyrgyz official pointed out:

> The question of elites is now the most important question in Kyrgyzstan. Everyone thinks that the problems of the South are problems of water and land, but above all it's a question about the elite.[13]

Further on in the chapter it will be discussed just how prophetic this warning proved to be.

The small elite that de facto ruled Kyrgyzstan, therefore, garnered much of the political and economic power for themselves to the detriment of any real democratic or free-market development. This was completely by design. It was thus a doubly dangerous system in that not only did the masses feel ignored but the majority of elites were also isolated and prevented from power. This was the foundation of the Kyrgyz political system from the very beginning and continued relatively unabated into the 21st century. This latter fact is best illustrated by three examples: the 2000 presidential election, the 2002 Ak-Sui shootings, and the 2005 parliamentary elections that culminated in the so-called 'Tulip Revolution' with Akaev fleeing from Kyrgyzstan.

The 2000 Presidential Elections

Before the results and discussion about the 2000 Kyrgyz presidential election can be written, it is necessary to provide a backstory more conspiratorial than most Hollywood screenwriters could provide. The man who would have been Akaev's main political rival in the 2000 race, Feliks Kulov, was once one of Akaev's most trusted political allies. He was ultimately barred from participating in the race but the manner in which he was denied is fascinating and telling.[14]

Kulov first came to fame by playing a significant role in ending the 1990 ethnic violence discussed earlier in Osh. He also provided strong military support to Akaev's regime during the fall of the Soviet Union. As a result of this Akaev named Kulov his first vice-president in 1992. He would also go on to serve as Minister of National Security. It was at this high point in his career that things began to unravel. While serving as National Security Minister Kulov was accused of participating in an aborted coup attempt, which he vehemently denied. At the time accusations were declared a curious theft took place: most of the equipment that had been allocated to a special task force set up by Kulov, and which included electronic eavesdropping devices, mysteriously disappeared.[15] With Akaev now horrendously suspicious of Kulov but with little to no direct evidence connecting him to either the coup or the theft, he tried to politically isolate him by demoting him to mayor of Bishkek in 1998. Kulov did not long stand for this dressing down and dramatically resigned as mayor of Bishkek in April 1999, writing an open letter to Akaev in which he declared, 'I cannot work further under your leadership, because with your connivance things are taking place in Kyrgyzstan which are incompatible with democracy and the rule of law'.[16] It was three months later that Kulov formed the Ar-Namys party to oppose Akaev in the 2000 elections.

By the time the election rolled around Kulov was already under indictment and in an army garrison court for the catch-all crime of 'abuse of power'. Since no verdict had been given in court regarding the charges, administration attempts to force Kulov's name from the ballot were initially unsuccessful. Then at the last hour they succeeded in pulling off a brilliantly Machiavellian maneuver: Kulov was refused registration for the ballot after he declined to sit for an examination to prove his fluency in the Kyrgyz language.[17] Kulov was therefore not allowed to participate in the race.[18] It was against the backdrop of this soap opera that the presidential race played out in 2000.

Out of 2,537,247 registered voters, supposedly 1,960,201 participated, an impressive 77 percent turnout. Askar Akaev won his 2000 presidential reelection bid in a landslide. The official vote tallies were as follows:

Table 4.1 2000 Kyrgyzstan Presidential Election Results

Askar Akaev:	1,460,201	74.49%
Omurbej Tekebaev	272,427	13.88%
Almaz Atembaev	117,658	6.00%
Melis Eshimkanov	21,260	1.08%

Across the board international organizations and foreign governmental agencies judged the elections to be rife with voter-rigging and voter fraud. These institutions included Human Rights Watch, the US Department of State and the Organization for Security and Cooperation in Europe. Not surprisingly the Central Electoral Committee in Bishkek declared the elections valid and in response both the State Department and the OSCE backed off from language that indicated there could be serious repercussions. Interestingly, the only major protests that transpired inside of Kyrgyzstan after Akaev's reelection occurred when protestors blocked the main north-south highway in Jalal-Abad province, which is in the Ferghana Valley.

As for Kulov, he was acquitted of the abuse of power court martial in August 2000 only to be arrested again and sentenced in January 2001 to seven years imprisonment on charges of 'abuse of office while serving as national security minister'.[19] In May 2002 Kulov was sentenced to *another* ten years imprisonment and just for good measure was barred from holding public office for a period of three years after the completion of his sentence. All of this was largely regarded by Kyrgyz society as Akaev pettiness: the president had come to detest Kulov and resented his potential as a political rival. Kulov would remain in jail until the Spring of 2005, when the 'Tulip Revolution' began to blossom and Kulov would once again find himself in the middle of political drama.

The 2002 Ak-Sui Shootings

In a move regarded by most local opposition groups as Akaev trying to gain personal favors from the Chinese government, Kyrgyzstan and China signed an odd settlement agreement over disputed territory along the far eastern edge of Kyrgyzstan. Despite vehement protests from Kyrgyz nationalists and parliamentary deputies, Akaev agreed to sign over more than 95,000 hectares to the Chinese government without any major explicit concessions from China.[20]

Azimbek Beknazarov, chairman of the parliament's Judicial and Legal Reform Committee, was especially vocal in how the agreement was ostensibly a unilateral concession made by the president that violated Kyrgyzstan's territorial integrity. On 8 January 2002 authorities charged Beknazarov with various 'violations in connection with the handling of a 1995 murder case while serving as a district prosecutor'.[21] These blatantly trumped-up charges caused several protests and hunger strikes across parliamentarian groups and civil society organizations. What happened next was best reported by the International Crisis Group:

> The government was shocked out of its complacency on 17 March 2002 as the protests started to grow out of control. Some demonstrators began to travel to the small town of Kerben to protest against Beknazarov's continued detention; others set off for Tash-Kumyr, a minor town on the main Bishkek-Osh road, with the aim of blocking the highway. A group of several hundred traveling to Kerben was stopped by police near the hamlet of Bospiek. The officials refused to let them travel further and in the ensuing confrontation police opened fire, killing four demonstrators.

> The news of the shootings traveled quickly and protestors in Kerben immediately surrounded the police station threatening to burn it down. Further clashes, in which one person died, ensued on March 18 in Kerben between police and protestors, whose numbers had now reached over 6,000. There was no real evidence to suggest that the demonstrators had firearms, despite attempts by the authorities to suggest otherwise. The violence only died down when it was announced that Beknazarov had been released, although he would still have to face trial at a later date.
>
> Akaev, according to people who met him about this time, seemed to have believed that the riots were the fault of the opposition, who were conspiring against him, and was reluctant to make any concessions. But finally, as rumors suggested that protestors were preparing to march to Bishkek, he was forced to act. The State Commission on the shootings in Ak-Sui finally published its report, after repeated delays, on 17 May 2002 and pointed to the arrest of Beknazarov as the main cause of the unrest.
>
> Under pressure…the president…dismissed three key political figures: Prime Minister Kurmanbek Bakiev, Head of the Presidential Administration Amanbek Karypkulov, and Minister of the Interior Temirbek Akmataliev. According to the constitution the resignation of the prime minister meant that the government was also obliged to resign.
>
> The dismissal of the government temporarily eased tensions and the blockade of the Osh-Bishkek road was lifted. On May 24, however, the trial of Beknazarov was resumed in Tokotogul. He was found guilty and given a one-year sentence. The judge ordered that he had already served long enough in remand so he was immediately released but the guilty verdict meant that he would automatically lose his mandate as a parliamentary deputy, something that his supporters refused to accept. They promised to protest further while Beknazarov appealed…On 28 June 2002 the Tokotogul court closed the case against Beknazarov, allowing him to retain his parliamentary deputy's mandate.[22]

There are a few observations to make with this example. First, the audacity of the government in the face of not only legitimate political opposition but also searing popular protest was incredible. The original charges against Beknazarov were conveniently created. The government initially refused to acknowledge his role in the emergence of protests and civil unrest. When the government's own official commission admitted Beknazarov's leading role, the 'sham trial' against him nonetheless continued. When the trial finally ended Beknazarov may have been relieved to be out of prison but the government still intended to silence his authority as a relevant political player by shackling him with a guilty verdict and ending his parliamentary eligibility.

Even when the entire affair was finally closed for good with Beknazarov basically in the same position as he was before the scandal, Akaev still was not yet ready to concede it a total wash. His dismissal of Prime Minister Bakiev forced the resignation of the entire government. Akaev promised to take into account the lessons learned from Ak-Sui when reappointing a new government but Akaev's 'coalition,' which involved only a small part of the opposition, turned out to consist mainly of middle-ranking bureaucrats that were still closely connected to the presidential family. There were no high posts for the opposition and there was little increase in southern Kyrgyz representation.[23] The only lessons learned from Ak-Sui, therefore, were twofold: First, it was ok to sacrifice the government for personal interest. Second, it was ok to make promises to the people only to then arrogantly ignore them without fear of consequences.

There would be consequences, however, and they would ironically involve not only former Prime Minister Bakiev but also former Vice-President Kulov. They would both ultimately benefit the most from the so-called 'Tulip Revolution' in March 2005. But before the Tulip Revolution is dealt with in greater detail, it is necessary to cover a few remaining points about the regime.

Clan Rivalries

Kyrgyz identity in both public and private life is largely determined by clan affiliation. There are basically three clan groupings in Kyrgyzstan, commonly referred to as 'wings'. You either belong to the 'right' wing, or Ong, the 'left' wing, or Sol, or the wing that is neither 'right' nor 'left', called Ichkilik.[24] The Ong wing includes seven clans from the north and the west. Each of the seven has consistently fought with the others for influence over the wing. The Buguu clan, for example, dominated the Ong wing historically, providing most of the first Kyrgyz administrators in the early Soviet period. Stalin's purges in the 1930s basically annihilated this leading clan and allowed another one, the Sarybagysh, to come to prominence.[25] Since Stalin the rule of the Sarybagysh clan has been uninterrupted. Akaev is a member of this clan.

The Sol wing contains a single clan, the Adygine, whose historical roots and present political power emanates from Southern Kyrgyzstan. The Ichkilik wing, which also has strong southern roots, is made up of numerous clans, some of whom are not even ethnically Kyrgyz but still claim Kyrgyz national identity.[26] If there was any significant change in this general clan structure under Akaev it was only in his concentrated effort to extend control of his own Sarybagysh clan over most of the key economic and political posts. This ultimately was reflected by the major ministries of finance, internal affairs and state security largely being filled by Akaev's clan.[27]

The internationally famous writer and Kyrgyz diplomat, Chingiz Aitmatov, has commented on the north-south dimension in Kyrgyz life, railing against the continued prominence of clan rivalries. Aitmatov has long called for 'partisans from both areas to set aside their sectional rivalries and work together on Kyrgyzstan's economic and political development'.[28] Fascinatingly, just as with Alexander Solzhenitsyn when he returned to Russia after decades in exile, Aitmatov has found the younger generations of Kyrgyz society not as willing to listen to words of reason. Most have simply called Aitmatov out of touch with the actual depth of resentment felt by the south towards the north and that he did not appreciate how much Akaev's policies were despised.

One of Akaev's more infamous policies was the appointing of political protégés to the Governor's post of Osh, one of the key cities of southern Kyrgyzstan. The previous five governors of Osh have all been northerners from Akaev's clan.[29] Most of Kyrgyzstan's prominent opposition leaders are all from the south, mainly from the Ichkilik clan. Most opposition bases of operation are located in the Osh, Jalal-Abad or Batken regions (all three regions are within the Ferghana Valley). Ultimately, it was clan life in Kyrgyzstan that resulted in further unrest after Ak-Sui. With a dismissed government and phony new appointments, it was clear to all in the south that Akaev had made no attempt to address the clan imbalance.[30]

Stoking the Fire: Islamic Persecution

Throughout the 1990s Central Asian regimes were relatively uniform in wanting to limit the scope and power of Islamic religious movements. The problem of course was that anti-Islamic strategy was so arbitrary and discriminatory that the only trend visible across the region since 2000 had been a subtle yet persistent increase in Islamic extremism. This work argues against underestimating that increase – examining as it does how states like Kyrgyzstan have stomped down on Islamic groups. Seeing how these groups have subsequently reacted and re-organized gives me pause to worry about one bitter political irony. Groups being harbored in the Ferghana Valley may in the end be ineffective for producing revolutions that overthrow *local* regimes but they may prove extremely effective for producing a bloody impact on *international* societies.

Despite all the crackdowns, new Islamic groups continue to emerge within the region. In Kyrgyzstan this was best exemplified by the creation of the Islamic Democratic Party of Kyrgyzstan (IDP). The IDP, based predominantly on the support of the southern poor, had aims to attract upwards of 100,000 members in order to challenge the secular Kyrgyz regime. Interestingly, many outside observers consider the potential impact of IDP to be minimal *because of the problems it will have registering as an official political party.*[31] Since IDP has a decided religious and confrontational angle in its ideology it is at the moment constrained by Kyrgyz law. In other words, it is dismissed because the political environment in-country is so repressive and discriminatory that legitimate means remain largely unavailable to them. So where must it go? Inevitably it will go underground. The one truism beyond debate is that the only surefire way to radicalize an Islamic group is to force it underground. This is the devil's dilemma facing all the Central Asian states and maddeningly underemphasized in American diplomacy: yes, radical groups legitimately participating in the political process can be a scary thing. Yet scarier still is if you force them out of legitimately participating. This has happened all over Central Asia and has been quietly consented to by American diplomatic circles.

It is naïve to think that US involvement in Afghanistan and Iraq, facilitated through access to air bases like the one at Manas in Kyrgyzstan, has not had an impact on the further radicalization of Islamic groups. The Hizb-ut-Tahrir (HT) has intensified its propaganda distribution in Kyrgyzstan since the invasion of Baghdad, with leaflets taking a decided anti-American stance:

> Let's rebel against the faithless...The war that [US President George] Bush started is a colonial war aimed at achieving hegemony and control, imposing influence and reshaping the region according to the new American standards.[32]

In Kyrgyzstan most Islamic groups, radical or not, were going to de facto violate either Article 297 or 299 of the Kyrgyz criminal code, in which efforts to 'bring about the forcible change of the constitutional system and attempts to foment national, racial and religious enmity' were forbidden.[33] Originally formulated to prevent ethnic violence, the legal article has been a tremendous weapon giving the executive administration near carte blanche freedom to viciously use its security services. This freedom has been used indiscriminately to monitor, eavesdrop, arrest without due cause and harass friends and family of those suspected of being members of banned groups. For example, during one spring month of 2003 in the Osh region alone law enforcement officials were conducting surveillance and intimidating more than 700 individuals with 'links to the IDP'.[34] It is obvious that such clumsy use of power has had the opposite intended effect within the valley: public protests have only increased and sympathy for radical Islamic groups has widened.

Clearly, as the Kyrgyz government clamped down on civil liberties and overextended its security mandate, it only succeeded in shrinking the political space for open and transparent discussion. Whether simply naïve or indifferent, the Kyrgyz government has always been wrong about the threat such Islamism represents. Tursunbai Baakir Uulu, the Kyrgyz government's first international ombudsman, claimed in 2003 that no Islamic movements represented a real threat to Kyrgyz security.[35] This lack of foresight seems to be universal: in this new world of global terror, too much time is spent concerned about old school threats against territorial integrity and theocratic revolution. While groups able to carry out such threats are indeed minimal in Central Asia this is in fact irrelevant: the new Islamic radical groups *are not making it a primary aim to accomplish such goals*. This is not their chief agenda.

After all, why aim for a revolution when civil unrest and pseudo-anarchy might take a government down for you? Symbolic terror, the apparent random and arbitrary infliction of pain and suffering on the civilian population, are powerful and constant threats to regimes. The focus on actual revolution is misplaced because it causes governments and scholars to consistently disregard more elemental, gradual threats, such as the fact that new radical Islamic groups are emerging while older ones are enlarging their numbers and moderates are growing increasingly more tolerant of radical views and philosophy. Radical Islamic propaganda is evolving beyond local politics to include transnational targets and objectives. But it is true that there have been no Islamic revolutions. Instead of being smugly overconfident at this so-called success of our western engagement we need to worry about one small thing: what if radicals aren't truly aiming for Islamic revolution in Central Asia?

This should be a cause for tremendous concern. Even if the United States could cynically care less about the safety of everyday Kyrgyz citizens it needs to understand the larger security issues Ferghana radicalism represents to its own interests: since 9/11 the trend toward smaller, more independent terror cells has increased while violent rhetoric has intensified. Such rigid compartmentalization, the isolation of one cell from another, is only efficient for one thing: training and conducting random acts of terror against civilian populations. Even the ethno-national geography of violence has gradually altered inside Ferghana since 9/11, evidenced by the capture of one Islamic radical in 2003 who admitted to participating in resistance campaigns in Afghanistan, Chechnya, Tajikistan, Kyrgyzstan and Uzbekistan.

That is the thing with radical transnational Islamic terrorists: they catch as catch can and go where ever the fight can be taken. Considering the ideal political geography of the valley and the political incompetence of local regimes, there was never reason to suspect that radicals in Central Asia would be happy exclusively aiming to kill Central Asians. The newest emerging facts indicate they would be happier hiding quietly in Ferghana while planning to kill us.

The Harvard Visit: Democracy the Kyrgyz Way

In 2004 President Akaev visited Harvard University and defended Bishkek's political record, saying that 'democratization efforts must correspond to historical practice in any given country'.[36] He said this as his country was heading into what were thought to be important parliamentary and presidential elections in 2005. Akaev by constitutional law could not stand for an additional term and so the atmosphere was tinged with suspicion: would Akaev finagle a way to get a legal right to stand for a third term? Would he seek to emulate Boris Yeltsin and 'handpick' a successor early so as to effectively make pre-election campaigns irrelevant? Akaev always denied these rumors but also was quick to reject notions of a 'universal formula' for democratic development.

He was passionate in his counter of foreign criticism, saying that 'western countries that expected a full-fledged democracy to take root quickly have forgotten their own experiences'.[37] Akaev praised his country for being the only one in the region to host both American and Russian military bases, with the American air base being at Manas since 2001 and the Russian base opening in Kant in 2003. The president proclaimed this positive balance was crucial in the common struggle all three have against Islamic radicalism.

Ultimately, the main gist of Akaev's talk focused on economic performance. While Bishkek initially courted foreign investment rather aggressively, expectations and results had remained negligible. Akaev explained why this necessitated a need for a new way of thinking:

> The orientation toward foreign models of economic development is internally defective, as is the import of reforms from the outside, even in the most delicate, velvet packaging...We decided that we would not attempt miracles and would develop our own model of development that takes into account our national specifics.[38]

This was Akaev's last major appearance on American soil before the events of Spring 2005. I include it because it accurately shows how susceptible he was to overthrow. Diplomatic rhetoric aside, his comments were exemplary in their ability to divert, sidetrack, cloud and ignore the most pressing issues that were facing his country. In the lead-up to parliamentary elections in March 2005 an intriguing and confusing chain of events took place throughout Kyrgyzstan that resulted in Akaev fleeing to Moscow and a new government being swept into power. It was thought that this was the moving eastward of the 'color' revolutions that had already caught on in Ukraine and Georgia. Washington initially welcomed the change warmly. A closer examination of events, people and policy behind this 'democratic revoliution', however, reveals slightly less revolutionary joy.

March 2005: The Tulip Revolution or the Criminal Coup?

Thomas Carothers wrote about 'semi-democracy,' the process whereby incomplete democratic transition may not necessarily be an intermediary stage to a greater liberation of politics.[39] States that have a dominant ruling elite and a weak opposition can actually maintain a sort of dysfunctional equilibrium where there is limited and ineffectual political pluralism while the entire system continues on unchanged in perpetuity. If popular support is low while a strong, corrupt class of business and political elites become strategically alligned, then the political system overall faces relatively little risk of overthrow. This was the situation in Kyrgyzstan come spring 2005. A detailed timeline follows in an attempt to piece together the various discombobulated strands of this 'revolution' that was meant to signal a positive change in Kyrgyz democracy.[40]

Unfortunately, this hope had already been smashed six months removed from the 'Tulip revolution'. Ironically, it seems that Akaev's 'semi-democracy' may now be best characterized as Bakiev's 'criminocracy'. In either case, the consequences to functional democracy and the global war against terror are not good.

- January 15: Before polling even began the Akaev government was attempting to discredit opposition candidates. This sparked civil unrest in early January when Roza Otunbaeva, a former Kyrgyz envoy to Britain and the United States, was prevented from participating in the elections on account of her 'continued long absences in residence outside of the country'.

- January 20: Akaev attempts to preempt popular protest by calling on his own supporters to 'resist provocateurs and exporters of velvet revolutions'. Fears of nepotism prove to be confirmed when the official list of candidates for seats in the parliament are announced: the President's 28-year old son, Aidar Akaev, 32-year old daughter, Bermet Akaeva, two relatives of first lady Mairam Akaeva and the son of Prime Minister Nikolai Tanaev are all registered to run.

- January 25–26: Presidential press secretary Abdil Segizbaev warned journalists that any attempt to start a Kyrgyz velvet revolution would only result in civil war and violence.

- February 3–5: Presdient Akaev addressed a crowd of 3,000 youths at a rally in Bishkek. He called on them to be 'immune to disease-causing pink, orange and yellow revolutionary viruses'. Meanwhile a counter protest forms at Victory Square, calling for free elections and for Akaev to step down before presidential elections set in the fall.

- February 21–24: Thousands began street protests in support of the numerous candidates who declared candidacies in the parliamentary elections but were subsequently disqualified from final lists. The largest printing press connected to opposition newspapers, the Independent Printing Press, had the electricity to its building inexplicably cut.

- February 27: The first round of voting in parliamentary elections ends inconclusively. Half the seats get pushed to a second round of voting in March. The majority of seats that were decided heavily favor pro-Akaev parties. Accusations of vote-rigging and vote-buying are rampant, with Russian media observers dubbing the election the 'wallet revolution'. Protests began to erupt across the country but especially in the Ferghana regions to the south. Governmental authorities, incredibly, lamely accused opposition figures of paying people to protest in the streets. Aidar Akaev, son of the President, won his seat with over 70 percent of the vote. Bermet, Akaev's daughter, went through to the second round with a clear lead overall. Azimbek Beknazarov, the controversial figure at the center of the Ak-Sui shootings, won a seat from the Ak-Sui region.

- March 13: The second round of voting in parliamentary elections was a clear 'victory' for President Akaev, amidst mounting accusation of fraud and irregularities. Three pro-presidential parties, Alga Kyrgyzstan, Adilet and the Democratic Party of Women and Youth took one quarter of all seats. Most of the non-affiliated independent candidates were expected to switch and 'find' allegiance to one of the three pro-government parties before the official parliamentary session began, ensuring Akaev an absolute majority. Two days after the vote the opposition launched a local congress, called a 'kurultai,' in the Ferghana city of Jalal-Abad. As many as 15,000 people attended the congress, which passed a resolution demanding that Akaev step down.

Map 4.1 Kyrgyzstan in Central Asia

- March 18–22: The south was now fully involved in numerous and large-scale protests. Protestors who had been merging on the governor's office in Osh the previous ten days finally broke in and took the building over. Special police units were sent in to both Osh and Jalal-Abad in order to storm the regional governmental buildings. In the ensuing clashes fourteen protestors and five police officers were severely injured. On March 20 both buildings were successfully seized by police, only to be countered by 40,000 people who had converged in Jalal-Abad. Within hours they had recaptured the buildings. Akaev continued to deny the relevance of events transpiring around him, rambling about 'foreign backers' financing the protests. Otunbaeva proclaimed 'it is not a takeover but rather the complete demoralization of the authorities. The government has collapsed by itself'. On March 22 protestors arrive in Bishkek as the government organizes a 10,000 strong pro-government rally to witness the swearing in of the 'new democratically-elected' parliament. Government negotiators begin talks with the Osh opposition while protest rallies are broken up by police in Bishkek.

- March 24–26: This first day marks the official start of the 'Tulip revolution' as a crowd of 1,000–2,000 opposition supporters stormed the main government building in Bishkek. By afternoon, the outgoing parliament convened with the Supreme Court and the Central Electoral Commission. Ishenbay Kadyrbekov, a leading opposition figure, was selected as the new speaker while Kurmanbek Bakiev, the one-time prime minister under Akaev and now staunch opponent, was named as interim prime minister. Two 'parallel' parliaments were formed as old and new legislatures signed a memorandum recognizing each other's authority for the following two weeks. On March 25 Akaev, now already in Moscow after having fled Kyrgyzstan, issued a statement on Russian television claiming his presidency was still intact. Bakiev appealed to the parliament, declaring 'if you trust me, and as far as I understand you do, give me the opportunity to form an executive body urgently. It will not stay in power forever, it won't even stay for several years, only for about three months'.[41] But by March 26 it was already clear the 'two parliaments' idea was not going to work as infighting had already begun between various factions. The Central Electoral Commission met and ruled that only the new parliament was in fact legitimate. The date for new presidential elections was moved up four months, to June 26. Feliks Kulov, the former vice-president and interior minister, is freed by opposition supporters from jail and is given a special interim designation as head of national security.

- March 27–29: The new single-chamber parliament went into session and elected Omurbek Tekebaev, one of the few opposition figures who won a seat in the original parliamentary elections, as speaker. His first act as speaker was to send a parliamentary delegation to Moscow to meet with Akaev to negotiate his resignation.

- April 4: Askar Akaev officially resigns his post as President of the Republic of Kyrgyzstan.

- April 8–11: Parliamentary deputies voted to strip Akaev of most of his rights as an ex-President, although they curiously leave his immunity from prosecution in place, and deprived his family of continuing rights to enjoy state support. Parliament formally accepted Akaev's resignation as head of state on April 11. A new date is set again for presidential elections, July 10.

This relatively peaceful 'revolution' was widely welcomed in Washington. It was hoped that events in Kyrgyzstan would have some wider significance for the region, an opportunity to push the 'Bush doctrine' of globalizing American values.[42] Many in the Bush White House even saw the doctrine as a major source of inspiration for Kyrgyzstan's popular revolt. While both Bakiev and Kulov were not new faces in Kyrgyz politics, they were 'known' ones to the White House and many believed the change from the increasingly irrational Akaev would only signal a reestablishment of close Kyrgyz-US relations.[43] Even though the government turnover did not wholly conform to democratic principles Washington was eager to assist the new government, downplaying concerns of legitimacy.[44]

In other words, Washington acted with the transitional Bakiev administration exactly as it acted for 15 years with the Akaev administration: take what it could get and be damned with the principles. This proved to be a fundamental error as the Kyrgyzstan revolution has not played out the way many anticipated. It was an even more egregious mistake considering most of these problems were readily apparent.

One day after Akaev fled Bishkek and Bakiev was named interim Prime Minister, some European-based Kyrgyz analysts were warning of possible problems. Murad Esenov, editor-in-chief of the Sweden-based 'Central Asia and the Caucasus' journal, said a power clash between Kulov and Bakiev would be inevitable.[45]

> As we know, Kulov represents the country's north, while Bakiev is a representative of the south…I believe they [both] turned against Akaev [despite their desire to remain in office]. It was Akaev who asked them to resign. If they continued to have authority, they would have still been in power. I don't think they were forced to resign because they were proponents of democracy. After they were moved away from power, they started to fight for it. I don't believe there were ideological disagreements between Bakiev, Kulov and Akaev.

Esenov would go on to worry that many 'new' opposition leaders seemed to be former high-ranking officials. This only indicated to him the absence of true democratic pluralism. Even more disturbing were the quickly changing reports about the forces that spurred on the revolution. It started to emerge that there may have been more than a passing interest from organized crime when it pertained to southern protests leading up to March 24.[46]

The Criminal Connection

Kyrgyzstan throughout the 1990s had always been a major transit corridor for Afghan drugs smuggled to Russia and on to Western Europe. Despite well-documented levels of corruption throughout the Akaev government, there was never any evidence linking the highest levels of Akaev's circle with these drug transfers. Plenty of lower level police would be involved in the reselling of confiscated drugs but this was minor and more a symptom of horrifically low pay. The 2005 revolution, however, indicates a new trend in Kyrgyz politics where organized crime has an especially privileged place.

Southern Kyrgyzstan's drug barons have for a long time been infamous for secretly forming their own paramilitary forces. These forces were usually 'hidden' under the guise of popular wrestling clubs or martial arts academies.[47] The work of Cornell and Swanstrom revealed that approximately 2,000 young sportsmen from these various clubs and academies were gathered, chauffeured and fed for twenty five days in order to storm state offices in Jalal-Abad, Osh and ultimately Bishkek.[48] This was confirmed by me independently, speaking with a political journalist in Bishkek who asked for anonymity and talked about the widespread drunkenness and odd 'athletic' body-types of the most violent protestors.[49]

Kyrgyz opposition was always traditionally disorganized because of Akaev interference. This made criminal involvement almost logical: substantial funds were lacking and structured organization was weak. This most likely accounted for the maddening indifference by the Akaev government leading up to the March events: in the past opposition was always hindered by these structural flaws to such a degree it was never able to reach its 'tipping point' to trigger a true coup. It seems in March, however, criminal leaders finally made the clear path they had always wanted to infiltrate and influence the opposition movement.

As to who was responsible for such a devil's bargain, there were very few legitimate choices. Bayaman Erkinbayev, a leading opposition figure who became a member of parliament in the spring events, also happened to be one of southern Kyrgyzstan's richest and most influential men. Erkinbayev controlled the majority of wrestling and martial arts clubs in southern Kyrgyzstan. In addition, acting interim President Kurmanbek Bakiev raised eyebrows in May 2005 when he appointed Dastan Sarygulov as state secretary. Sarygulov was earlier involved in gold mining business scandals that involved the mysterious disappearance of vast amounts of Kyrgyz gold revenues.[50]

Process of elimination can be a useful tool in political analysis. Three unfortunate events, one happening before the new presidential elections and two happening afterward, seemed to indicate who the true culprit was. On April 5 Colonel Uran Aliev was gunned down outside of his house in Bishkek by unknown assailants. Aliev was the head of the Osh regional department for the Interior Ministry and was charged with combating state crimes.[51] It was believed at the time of his death Aliev was on the verge of exposing high-level corruption that extended into the *new interim administration*.

On September 19 Bakiev dismissed Azimbek Beknazarov as prosecutor general.[52] Beknazarov had also been aggressively pursuing corruption cases, including several that went beyond former President Akaev and reached into the new administration. Allies of Bakiev expressed discontent with Beknazarov's 'maverick style,' claiming he had overstepped his bounds and had an inflated sense of the importance of the prosecutor general's office.[53] At a September 20 news conference, Beknazarov laid the finger of blame firmly on the door of President Bakiev. He alleged that corruption was rampant within the new administration and, more damagingly, that Bakiev himself didn't really mind. His official statement read, 'Bakiev sacrificed Beknazarov to the interests of criminality. Incumbent authorities are not interested in the struggle against corruption. Organized criminal elements have begun to openly cooperate with officials'.[54] Other local analysts felt the maneuver killed two birds with one stone: not only would the investigations end but Bakiev appointed a close friend and ally, Busurmankul Tabaldiyev, as prosecutor-general in Beknazarov's place.

Two days later on September 21 Erkinbayev was assassinated as he was returning to his home in Bishkek. While it is impossible, considering the nature of Erkinbayev's business interests and the nature of Kyrgyz politics, to determine the exact reason and culprits involved experts in post-Soviet studies would not be shocked to hear that all such affairs are usually inextricably linked. Yes, there was a criminal element to his assassination but the entire 'Tulip revolution' was in fact infused by an air of criminality from the start. Erkinbayev's death caused great consternation within parliament, leading some deputies to consider the rise in criminality under Bakiev to rival and even surpass anything that existed during the Akaev regime. One parliament deputy, Kabai Karabekov, compared the presidential administration to a 'mafioso structure'.[55]

So much for a revolution meant to signal happier days on the horizon, facilitated by a Bakiev administration that was for 'more democratization and a battle with corruption'.[56] This last statement must have been particularly ironic to the forces apparently behind Bakiev's landslide electoral victory in July. He won nearly 89 percent of the vote with a supposed 75 percent voter turnout. Even his electoral victory was not without incident as the OSCE reported 'serious breeches of transparency safeguards' in regards to vote-counting and results-tabulation.[57] Nearly a third of all polling stations monitored by international observers assessed counting procedures as BAD or VERY BAD. It does not require a giant leap of logic to ascertain what transpired at the polling stations where there were no international observers. A simple example involved the Leylek election district, where officials declared 650 votes were cast for Bakiev during a fifty minute span during the lunch hour. No observers were able to verify anywhere close to that amount of activity at any point in the day, let alone during 'lunchtime'.[58]

In little more than a calendar year, the Bakiev administration has been marked by dubious appointments, the killing of one lesser-known official investigating corruption and the dismissal of a better-known official also investigating corruption and criminality. Incredibly, recall that Beknazarov was allowed to become a member of the present parliament because the Akaev government was *afraid* what his removal might do in regards to civil unrest. Repressing Beknazarov had already indirectly caused the Ak-Sui shootings. Akaev didn't want to risk another civil disturbance. How ironic that the new government, meant to do away with Akaev arbitrariness, was unafraid to do what Akaev feared. In addition to these disturbing discrepancies, a major political rival was assassinated and vote counts within the presidential election were rigged to create a fake mandate.

Neutering the Constitutional Court

As a member of the anti-Akaev opposition, Bakiev was always a strong supporter of amending the constitution. It was essential, according to Bakiev, to reinstitute the system of checks and balances into the Kyrgyz Constitution so that the legislative and executive branches would be in more rational equilibrium with each other.[59] Not surprisingly, Bakiev's taste for curbing executive authority noticeably lessened once inaugurated into office. One of the small sub-texts before the March

24 leadership change was a Constitutional Council charged with drafting amendments to the constitution to reign in Akaev. That council's work became bogged down after Bakiev came into office, no thanks in part to his repeated reshuffling of the council in order to 'reinvigorate' their work.[60] If there was any doubt as to the ultimate fate of the council under a new regime it was ended when Bakiev placed *himself* as chair of the council, a move that was actually supported by Parliament Speaker Tekebayev.

In other words, the institution in Kyrgyzstan meant to review and recommend changes that would limit and lessen the power of the president was being chaired by the president. This obvious and ridiculous conflict of interest was then supported by the parliament, which stood to gain power were the council to actually do its job. Clearly, checks and balances were also not a favorite institution of Bakiev.

Tekebayev intimated that Bakiev's landslide election win over the summer gave him a popular mandate that simply had to be respected.[61] Suddenly the inexplicable vote-rigging in an election that was already a foregone conclusion made sense: if the goal was not just to win but to win in such a way so as to create a mandate for limiting criticisms, then the strategy was incredibly successful. Unfortunately, it also undoubtedly signaled a continued backsliding of democracy in Kyrgyzstan.

This change in the course of executive 'constraint' was then followed up in November by a packet of *different* constitutional amendments aimed at 'reforming' the judiciary branch. Incredibly, the main amendment dominating the packet was a plan to abolish the Constitutional Court, which by law was the supreme arbiter on matters relating to the affairs of state and fundamental rights.[62] The amendments were approved in mid-November by a 289–member Constitutional Conference. A 'public' debate on the amendments had to officially take place after which President Bakiev would be able to order them into law. Daniyar Narynbaev, the president's official representative to the parliament, explained succinctly:

> Our small republic does not need so many higher courts. All efforts can be united in a single court. It's more an organizational decision than a legal one.[63]

Akaev was criticized in this chapter for his inexplicable inability to understand the depth of protest against his various corrupt maneuvers. Arguably this Constitutional Court maneuver from Bakiev tops any Akaev move in audacity. Chairing the council was audacious enough but then moving quickly to reform the judiciary branch so that the highest court in the country would no longer have a jurisdictional say over constitutional reform was beyond audacity. While the presidential administration argued this was not so much a legal change than an administrative one, the head of the Constitutional Court, Cholpon Baekova, disagreed:

> Former president Akaev did not spare the Constitutional Court either, and attempts to abolish it have been going on since 1994. At that time they tried to reduce the court's functions and later to raise a supervisory body above it. And

> the supporters of that scheme were the same experts who've been involved in developing the new version of the Constitution. The same process of making the Constitutional Court docile and obedient is continuing.[64]

The only thing that remains to be seen is if Bakiev can be successful to a larger degree than his predecessor. If he is, then the 'Tulip Revolution' is mutating a semi-democracy into one even less recognizable to Western standards.

Masking Media Control

This general 'reform' trend continued within the realm of freedom of speech. The draft Law on Mass Media came under criticism for containing 'illegitimate content restrictions' and being plagued overall by phraseology so vague that restrictions could be invoked over almost anything.[65] For example, a restriction was included on printing 'unprintable expressions'. Even more disturbing has been the dashed hopes of many media outlets following the March events. In the lead up to the parliamentary elections recall that Akaev had succeeded in garnering state control over the vast majority of media sources. It was considered obvious that the new government would begin initiatives to unshackle them from state control. Apparently old habits die hard.

In mid-October the government announced it would re-privatize KOORT, a television station once controlled by the Akaev family and a source of media criticism against Bakiev since his election.[66] *Vechernii Bishkek*, a leading opposition newspaper during the Akaev regime, was returned to Alexander Kim only after he became closely aligned with Bakiev. Another famous opposition source of media under Akaev, television station NTRK, has mysteriously become a huge fan of all Bakiev policies.[67]

All of this, taken in total, is not perhaps such a surprise politically. After all, Russia has been teetering and tottering along its own compromised democratic path even despite a change in the presidency. Most of the other former Soviet Republics today also cannot be considered true democracies. Why this is so important in Kyrgyzstan is because of the Ferghana Valley and the potential fatal consequences compromised democracy can have on the global war against terror.

Conclusion – Criminocracies and Their Impact on Rising Terror

There is no doubt that Akaev's 'managed democracy' had become burdened by its own internal contradictions and blatant disregard for freedom and liberty. This chapter documented how such disregard and contradiction was advantageous to the rise of a new kind of Islamic fundamentalism in the Ferghana Valley: one that was transnational in structure and international in its aims. If anything, the trends emerging from the Bakiev regime are even more dangerous. Managed democracy is inefficient and incompetent. Criminocracy actively undermines democracy and gives rise to a criminalization of politics that can only benefit terror groups as they seek to carve out a stable and benign operating space.

The American government held out hope with the removal of Akaev from power but all in all its stance toward Bakiev has changed little even as the various transgressions and disturbing trends have come to light. This is not because American analysts are horrific at understanding the subtleties of Kyrgyz politics. It is because the long-term goals essential to fighting terror (building functional democracy embodied by the principles of equality, liberty, justice transparency and freedom) simply never win the internal policy debate taking place at the White House. The short-term security strategists presently dominate White House decision-making and are simply too powerful and too unwilling to entertain alternative ideas.

The academic community is not innocent in this either. Time and again we have taken diplomatic proclamations from Washington as legitimate reflections of policy, even when empirical reality shows us something different. Before 9/11 it could be argued this was just an intellectual exercise criticizing the innate hypocrisy of modern diplomacy. In a post-9/11 world, however, this lack of scholarly attention leads to pain and suffering and for the first time that pain and suffering does not involve only the degradation of little-known peoples in little-known states like Kyrgyzstan. Post-9/11 it will involve America and include Americans. Hypocrisy in diplomacy, compounded by gullibility in scholarship, has simply become too costly. Kyrgyzstan's 'criminal coup' is therefore simply an allegory signifying the wider problem addressed in this piece.

What will be the consequences for Kyrgyzstan now that it is under the leadership of a president who is already referred to locally under the nickname of 'President BAKS?' This is a reference to Bakiev, Kurmanbek Salievich, the full name of the president and, in an amazing coincidence, identical with the slang term for AMERICAN DOLLARS, an obvious reference to Bakiev's criminal connections and corrupt policies. Local political analysts in Kyrgyzstan already openly discuss the fear of the 'Afghanization of Kyrgyzstan,' pointing out how some provinces within the republic are already not subordinating themselves to the federal structure, denying any real allegiance to the center.

Where these Kyrgyz analysts are mistaken, however, is in the fear that this makes them vulnerable to renegade Taliban escaping north to establish a new regime in Kyrgyzstan. As discussed in this chapter, the problem with 'President Bucks' is not how susceptible the republic is to Islamic theocratic revolution but how ideal the state has become as a hiding place for all those radical groups who wish to destroy the West. There was no greater place in the world for this location than Afghanistan before the United States removed the Taliban. Now, in regards to geographical ease of proximity and environmental accessibility, I can see no better replacement than the Ferghana Valley and the three states who share it. The 'Tulip Revolution' made no change against this judgment and in some ways appears to be making it more factual. The most damning thing of all, however, is that the United States seems ignorant to the ultimate consequences and danger such reality represents to us.

1 International Crisis Group, 'Kyrgyzstan's Political Crisis: An Exit Strategy', ICG Asia Report, No. 37, 19 August 2002.

2 International Crisis Group, 'Kyrgyzstan's Political Crisis: An Exit Strategy', ICG Asia Report, No. 37, 19 August 2002.

3 International Crisis Group, 'Kyrgyzstan's Political Crisis: An Exit Strategy', ICG Asia Report, No. 37, 19 August 2002.

4 Alisher Khamidov, Eurasianet, 'Kyrgyzstan: Organized Opposition and Civil Unrest', http://www.eurasianet.org/departments/rights/articles/eav121602.shtml, accessed 30 October 2003.

5 Alisher Khamidov, Eurasianet, 'Kyrgyzstan: Organized Opposition and Civil Unrest', http://www.eurasianet.org/departments/rights/articles/eav121602.shtml, accessed 30 October 2003.

6 Alisher Khamidov, Eurasianet, 'Kyrgyzstan: Organized Opposition and Civil Unrest', http://www.eurasianet.org/departments/rights/articles/eav121602.shtml, accessed 30 October 2003.

7 Alisher Khamidov, Eurasianet, 'Kyrgyzstan: Organized Opposition and Civil Unrest', http://www.eurasianet.org/departments/rights/articles/eav121602.shtml, accessed 30 October 2003.

8 Alisher Khamidov, Eurasianet, 'Kyrgyzstan: Organized Opposition and Civil Unrest', http://www.eurasianet.org/departments/rights/articles/eav121602.shtml, accessed 30 October 2003.

9 Alisher Khamidov, Eurasianet, 'Kyrgyzstan: Organized Opposition and Civil Unrest', http://www.eurasianet.org/departments/rights/articles/eav121602.shtml, accessed 30 October 2003.

10 International Crisis Group, 'Kyrgyzstan's Political Crisis: An Exit Strategy', ICG Asia Report, No. 37, 19 August 2002.

11 International Crisis Group, 'Kyrgyzstan's Political Crisis: An Exit Strategy', ICG Asia Report, No. 37, 19 August 2002.

12 International Crisis Group, 'Kyrgyzstan's Political Crisis: An Exit Strategy', ICG Asia Report, No. 37, 19 August 2002.

13 International Crisis Group Interview, Jantoro Satybaldiev, Mayor of Osh, Kyrgyzstan, 21 June 2002.

14 Alisher Khamidov, Eurasianet, 'Kyrgyzstan: Organized Opposition and Civil Unrest', http://www.eurasianet.org/departments/rights/articles/eav121602.shtml, accessed 30 October 2003.

15 Alisher Khamidov, Eurasianet, 'Kyrgyzstan: Organized Opposition and Civil Unrest', http://www.eurasianet.org/departments/rights/articles/eav121602.shtml, accessed 30 October 2003.

16 Radio Free Europe/Radio Liberty, 'Profile: Feliks Kulov', http://www.rferl.org/featuresarticle/2005/03/ec9e9641–1632–4981–a435–21505d93a517.html, accessed 16 November 2003.

17 Radio Free Europe/Radio Liberty, 'Profile: Feliks Kulov', http://www.rferl.org/featuresarticle/2005/03/ec9e9641–1632–4981–a435–21505d93a517.html, accessed 16 November 2003.

[18] Eurasianet, 'Kyrgyzstan Local Press Digest', Digest #12, 1–17 July 2000, http://www.eurasianet.org/resource/kyrgyzstan/press_digest/Ky071800.shtml, accessed 24 October 2003.

[19] Radio Free Europe/Radio Liberty, 'Profile: Feliks Kulov', http://www.rferl.org/featuresarticle/2005/03/ec9e9641–1632–4981–a435–21505d93a517.html, accessed 16 November 2003.

[20] Alisher Khamidov, Eurasianet, 'Kyrgyzstan: Organized Opposition and Civil Unrest', http://www.eurasianet.org/departments/rights/articles/eav121602.shtml, accessed 30 October 2003.

[21] Alisher Khamidov, Eurasianet, 'Kyrgyzstan: Organized Opposition and Civil Unrest', http://www.eurasianet.org/departments/rights/articles/eav121602.shtml, accessed 30 October 2003.

[22] International Crisis Group, 'Kyrgyzstan's Political Crisis: An Exit Strategy', ICG Asia Report, No. 37, 19 August 2002.

[23] International Crisis Group, 'Kyrgyzstan's Political Crisis: An Exit Strategy', ICG Asia Report, No. 37, 19 August 2002.

[24] Alisher Khamidov, Eurasianet, 'Kyrgyzstan's Unrest Linked to Clan Rivalries', http://www.eurasianet.org/departments/insight/articles/eav060502.shtm, accessed 4 January 2004.

[25] Alisher Khamidov, Eurasianet, 'Kyrgyzstan's Unrest Linked to Clan Rivalries', http://www.eurasianet.org/departments/insight/articles/eav060502.shtm, accessed 4 January 2004.

[26] Alisher Khamidov, Eurasianet, 'Kyrgyzstan's Unrest Linked to Clan Rivalries', http://www.eurasianet.org/departments/insight/articles/eav060502.shtm, accessed 4 January 2004.

[27] Alisher Khamidov, Eurasianet, 'Kyrgyzstan's Unrest Linked to Clan Rivalries', http://www.eurasianet.org/departments/insight/articles/eav060502.shtm, accessed 4 January 2004.

[28] Alisher Khamidov, Eurasianet, 'Kyrgyzstan's Unrest Linked to Clan Rivalries', http://www.eurasianet.org/departments/insight/articles/eav060502.shtm, accessed 4 January 2004.

[29] Alisher Khamidov, Eurasianet, 'Kyrgyzstan's Unrest Linked to Clan Rivalries', http://www.eurasianet.org/departments/insight/articles/eav060502.shtm, accessed 4 January 2004.

[30] Alisher Khamidov, Eurasianet, 'Kyrgyzstan's Unrest Linked to Clan Rivalries', http://www.eurasianet.org/departments/insight/articles/eav060502.shtm, accessed 4 January 2004.

[31] David Montgomery, Euarsianet, 'Religious Persecution and Extremism Fosters Discontent in Central Asia', http://www.eurasianet.org/departments/rights/articles/pp083002.shtml, accessed 15 July 2003.

[32] Alisher Khadimov, Eurasianet, 'Islamic Radical Group Intensifies Underground Activity in Kyrgyzstan', http://www.eurasinet.org/departments/rights/articles/eav041403.shtml, accessed 18 October 2004.

[33] Alisher Khadimov, Eurasianet, 'Islamic Radical Group Intensifies Underground Activity in Kyrgyzstan', http://www.eurasinet.org/departments/rights/articles/eav041403.shtml, accessed 18 October 2004.

[34] Alisher Khadimov, Eurasianet, 'Islamic Radical Group Intensifies Underground Activity in Kyrgyzstan', http://www.eurasinet.org/departments/rights/articles/eav041403.shtml, accessed 18 October 2004.

[35] Alisher Khadimov, Eurasianet, 'Islamic Radical Group Intensifies Underground Activity in Kyrgyzstan', http://www.eurasinet.org/departments/rights/articles/eav041403.shtml, accessed 18 October 2004.

[36] Erdin Beshimov, Eurasianet, 'Akaev: We do Democracy the Kyrgyz Way', http://www.eurasianet.org/departments/insight/articles/eav101504.shtml, accessed 23 March 2005.

[37] Erdin Beshimov, Eurasianet, 'Akaev: We do Democracy the Kyrgyz Way', http://www.eurasianet.org/departments/insight/articles/eav101504.shtml, accessed 23 March 2005.

[38] Erdin Beshimov, Eurasianet, 'Akaev: We do Democracy the Kyrgyz Way', http://www.eurasianet.org/departments/insight/articles/eav101504.shtml, accessed 23 March 2005.

[39] Thomas Carothers, *Aiding Democracy Abroad: The Learning Curve*, Carnegie Endowment for International Peace, Washington DC, 1999.

[40] The following timeline is culled and organized from International War and Peace Reporting, http://www.iwpr.net/index.pl?centasia_kyrgyzrev.html, accessed 14 November 2005.

[41] Gulnoza Saidazimova, Eaurasianet, 'Kyrgyzstan: New Leadership Drawn from the Old Guard', http://www.eurasianet.org/departments/insight/articles/pp032505a.shtml, accessed 15 November 2005.

[42] Ariel Cohen, Eurasianet, 'Bush Administration Welcomes the Kyrgyz Revolution', http://www.eurasianet.org/departments/insight/articles/eav022905.shtml, accessed 24 May 2005.

[43] Ariel Cohen, Eurasianet, 'Bush Administration Welcomes the Kyrgyz Revolution', http://www.eurasianet.org/departments/insight/articles/eav022905.shtml, accessed 24 May 2005.

[44] Ariel Cohen, Eurasianet, 'Bush Administration Welcomes the Kyrgyz Revolution', http://www.eurasianet.org/departments/insight/articles/eav022905.shtml, accessed 24 May 2005.

[45] Gulnoza Saidazimova, Eaurasianet, 'Kyrgyzstan: New Leadership Drawn from the Old Guard', http://www.eurasianet.org/departments/insight/articles/pp032505a.shtml, accessed 15 November 2005.

[46] Svante Cornell and Niklas Swanstrom, Central Asia-Caucasus Analyst, 'Kyrgyzstan's Revolution: Poppies or Tulips?', http://www.cacianalyst.org/view_article.php?articleid=3323, accessed 27 June 2005.

[47] Svante Cornell and Niklas Swanstrom, Central Asia-Caucasus Analyst, 'Kyrgyzstan's Revolution: Poppies or Tulips?', http://www.cacianalyst.org/view_article.php?articleid=3323, accessed 27 June 2005.

[48] Svante Cornell and Niklas Swanstrom, Central Asia-Caucasus Analyst, 'Kyrgyzstan's Revolution: Poppies or Tulips?', http://www.cacianalyst.org/view_article.php?articleid=3323, accessed 27 June 2005.

[49] Interview with 'Alima N.', Bishkek, Kyrgyzstan, 5 December 2005.

[50] Svante Cornell and Niklas Swanstrom, Central Asia-Caucasus Analyst, 'Kyrgyzstan's Revolution: Poppies or Tulips?', http://www.cacianalyst.org/view_article.php?articleid=3323, accessed 27 June 2005.

[51] Svante Cornell and Niklas Swanstrom, Central Asia-Caucasus Analyst, 'Kyrgyzstan's Revolution: Poppies or Tulips?', http://www.cacianalyst.org/view_article.php?articleid=3323, accessed 27 June 2005.

[52] Eurasianet, 'Kyrgyzstan's Revolution at Risk', http://www.eurasianet.org/departments/insight/articles/eav092605.shtml, accessed 18 October 2005.

[53] Eurasianet, 'Kyrgyzstan's Revolution at Risk', http://www.eurasianet.org/departments/insight/articles/eav092605.shtml, accessed 18 October 2005.

[54] Eurasianet, 'Kyrgyzstan's Revolution at Risk', http://www.eurasianet.org/departments/insight/articles/eav092605.shtml, accessed 18 October 2005.

[55] Eurasianet, 'Kyrgyzstan's Revolution at Risk', http://www.eurasianet.org/departments/insight/articles/eav092605.shtml, accessed 18 October 2005.

[56] Eurasianet, 'Bakiev Wins Landslide in Kyrgyzstan, Courts Controversy with United States', http://www.eurasianet.org/departments/insight/articles/eav071105.shtml, accessed 15 August 2005.

[57] Eurasianet, 'Bakiev Wins Landslide in Kyrgyzstan, Courts Controversy with United States', http://www.eurasianet.org/departments/insight/articles/eav071105.shtml, accessed 15 August 2005.

[58] Eurasianet, 'Bakiev Wins Landslide in Kyrgyzstan, Courts Controversy with United States', http://www.eurasianet.org/departments/insight/articles/eav071105.shtml, accessed 15 August 2005.

[59] Eurasianet, 'Kyrgyzstan: Is the President Losing Interest in Constitutional Reform?', http://www.eurasianet.org/departments/insight/articles/eav082505.shtml, accessed 23 September 2005.

[60] Eurasianet, 'Kyrgyzstan: Is the President Losing Interest in Constitutional Reform?', http://www.eurasianet.org/departments/insight/articles/eav082505.shtml, accessed 23 September 2005.

[61] Eurasianet, 'Kyrgyzstan: Is the President Losing Interest in Constitutional Reform?', http://www.eurasianet.org/departments/insight/articles/eav082505.shtml, accessed 23 September 2005.

[62] Eurasianet, 'Kyrgyzstan: Is the President Losing Interest in Constitutional Reform?', http://www.eurasianet.org/departments/insight/articles/eav082505.shtml, accessed 23 September 2005.

[63] Eurasianet, 'Kyrgyzstan: Is the President Losing Interest in Constitutional Reform?', http://www.eurasianet.org/departments/insight/articles/eav082505.shtml, accessed 23 September 2005.

[64] Eurasianet, 'Kyrgyzstan: Is the President Losing Interest in Constitutional Reform?', http://www.eurasianet.org/departments/insight/articles/eav082505.shtml, accessed 23 September 2005.

65 Eurasianet, 'Promoting Media Freedom in Kyrgyzstan Proving More Difficult than Originally Anticipated', http://www.eurasianet.org/departments/insight/articles/eav110905.shtml, accessed 06 December 2005.

66 Eurasianet, 'Promoting Media Freedom in Kyrgyzstan Proving More Difficult than Originally Anticipated', http://www.eurasianet.org/departments/insight/articles/eav110905.shtml, accessed 06 December 2005.

67 Eurasianet, 'Promoting Media Freedom in Kyrgyzstan Proving More Difficult than Originally Anticipated', http://www.eurasianet.org/departments/insight/articles/eav110905.shtml, accessed 06 December 2005.

Chapter 5

Tajikistan: Coalition via Spin-Control

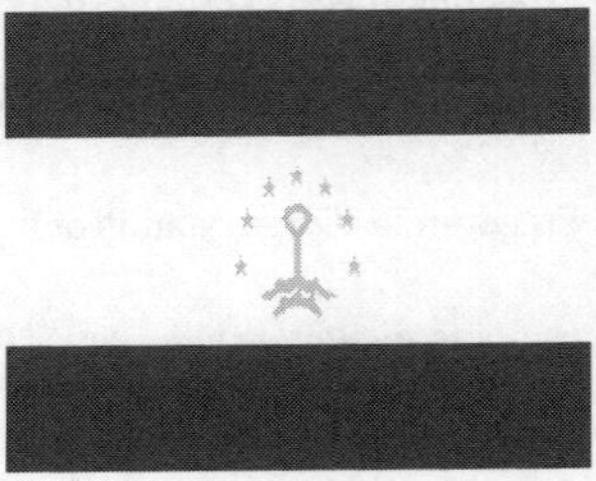

Tajikistan was the only state in Central Asia to suffer an actual violent civil war, from 1992 to 1997. It was also however the only state in Central Asia to resolve such violence through negotiated reconciliation that included a 'quota system' for the opposition in government. This opposition was captained by the Islamic Renaissance Party (IRP). This chapter will show how superficial and irrelevant this coalition was and is: the IRP has been a compromised party almost from the signing of the Peace Accord in 1997. Ideologically, theologically and legally its governing authority was and is a symbolic formality. The IRP is irrelevant as a model of true power for Islamists. At the same time the prevailing winds in Tajikistan, exacerbated by the cozier relationship developed with the United States since 9/11, blow ever more strongly to the side of repression and political domination. The overall tendency of the US to use 'Wonka vision' democracy reigns supreme here: it deemphasizes obvious problems to hail the state's unique status as the only regime with a legal Islamic party. This diplomatic blindness carries grave dangers as it is perceived as tacit approval of subjugation by radicalized Islamists.

The Foundations of the Tajik Civil War

Tajikistan was the only Soviet Central Asian state to react to Gorbachev's *perestroika* by creating numerous and diverse socio-political organizations and parties.[1] There were three main groups: Rastokhez, which was comprised largely of Tajik intelligentsia, the Democratic Party of Tajikistan (DPT), supported heavily by non-Tajik peoples in-country, and the Islamic Renaissance Party (IRP), which was especially popular across the rural and southern districts and was led by Said Abdullo Nuri and Hoja Akbar Turandzhonzoda. The various parties had a common unifying theme in that all wanted a democratic, progressive and staunchly anti-communist system.[2] Explaining how such a progressive start ultimately turned into a civil war reveals an effort to destroy the 'Islamic-ness' of the IRP and questions the commitment of President Rakhmonov to a true coalition.

The opposition of Tajikistan maintained a highly secular vision of anti-communism right from the beginning. Surprisingly, even the Islamic groups were not especially overt in their religiosity: overtones were rather moderate and internal attempts to make them more intense found little support within the Islamic groups themselves. For example, Davlat Khudonazarov, who would later stand as a presidential candidate for the democratic-Islamic opposition, said, 'a lot of what Islam has brought into the national mentality of the people of Central Asia undoubtedly hinders our spiritual and social development'.[3] Thus motivational forces informing early Tajik opposition movements were first anti-communist, then democratic and finally Islamic. This puts Tajikistan as a model for secular-Islamic coalition under question: Islam was the poor third cousin in Tajikistani politics.

From the beginning the republican government in Tajikistan was willing to play the 'Muslim card' to its benefit. In 1991, in response to massive opposition protests in Dushanbe, officials circulated rumors that Islamic protestors were demanding the formation of an Islamic republic. In response opposition leaders actually declared that, 'the creation of an Islamic republic goes against the democratic principles of Tajikistan development'.[4] It is one thing to counter rumors that will give worry to an ill-informed public. It is another to counter them by basically *confirming* the underlying prejudice of the government's propaganda: the IRP dealt with criticism of Islam by declaring basically Islam incompatible with democracy.

The August Coup in Russia as Instigator of Clan Politics in Tajikistan

The August 1991 coup attempt against Soviet leader Gorbachev increased tensions within Tajikistan. The First Secretary of the Central Committee of the Communist Party of Tajikistan Kakhar Makhkamov was secretly taped during the coup backing the State Emergency Committee in Moscow. This tape made its way to Gorbachev insiders soon after, resulting in Makhamov's resignation and replacement by Kadriddin Aslonov, the Chairman of the Tajik Supreme Soviet.[5] Aslonov called for the cessation of all Communist party activities on Tajik territory and nationalized all of the party's property. Although such maneuvers seemed to be a validation of burgeoning Tajik sovereignty it in fact exacerbated clan tensions as they all rushed in to fill the post-Communist power vacuum.[6]

There was no basis for formal constitutionalism in Tajikistan and the atmosphere was tinged with the potential for violence. This power void was quickly filled by criminal groups united along regional or clan lines. Incredibly, these clan and criminal groups gradually began mutating into Tajik battalions as the criminal leaders became de facto 'field commanders' that were ultimately incorporated into the government soon after the first presidential elections in 1991, which were largely rigged by the disintegrating Soviet powers in Moscow.[7]

Nine candidates contested for the presidency, with the DPT, IRP and Rastokhez forming a semi-alliance and nominating Davlat Khudonazarov. Former Communist Party First Secretary Rakhmon Nabiev, fully supported by Gorbachev, won the election despite accusations of vote-rigging and voter fraud.[8] In May Nabiev used emergency powers to form a 'presidential guard' personally loyal to him and aimed to disrupt the building opposition movement. Only the intervention of the Russian Army's 201st division stabilized the situation with a weakened Nabiev agreeing to a coalition 'government of national reconciliation'. He remained president but allocated a third of all ministerial posts to opposition parties.[9]

The declaration of national reconciliation and the supposed formation of a coalition government did not lessen tension or violence throughout the state, however. By September the militant opposition 'Youth of Dushanbe' took Nabiev hostage at gunpoint and ultimately forced him to resign. Akbarsho Iskandarov was made acting President while Abdumalik Abdullajanov was named acting Prime Minister.[10] Within a month after the Nabiev hostage situation a new power figure emerged as Tajik politics began to disintegrate.

Uzbek and Russian Interference and the Emergence of Emomali Rakhmonov

Uzbek President Islam Karimov was always concerned about Islamic spillover into his country. By the time of the Nabiev hostage situation Karimov was already beginning to sour on the former party first secretary. He chastised Nabiev's 'weakness' when the government of national reconciliation was obligated to include numerous Islamists.[11] From mid-November to early December 1991 the Tajik Supreme Soviet held its 16th Congress, where Nabiev officially resigned. The security situation was considered so dangerous that armed Uzbek military units safeguarded the Congress.[12] Emomali Rakhmonov, a relatively unknown official from the Kulyab clan who had previously been a field commander for the People's Front, was chosen to replace Nabiev. He immediately formed a government comprised mostly of Kulyabi and Leninabadi clan members and began overturning all of the government of national reconciliation initiatives.[13]

Even though Rakhmonov promised not to antagonize the opposition a full-scale war erupted soon after his ascendance. The consequences were devastating to Tajikistan: by the end of 1992 approximately 50,000 people were killed, about 100,000 fled to Afghanistan and Kyrgyzstan, over 600,000 civilians were displaced inside the country and more than seven billion dollars of damage imploded state infrastructure.[14] Once the Rakhmonov government's forces finally secured Dushanbe on 10 December 1992, they did so riding in Uzbek tanks and backed with air sorties flown by the Uzbek air force.[15]

By the beginning of 1993 Russia aimed to be recognized as the primary 'peacekeeper' across the post-Soviet expanse. This turned out to be more than it had bargained for, however, as the Tajik opposition used the mountains of Afghanistan and Kyrgyzstan to conduct hit-and-run operations against the Russians. These would prove remarkably effective at sapping Russian resolve. When Russia began suffering significant casualties in the summer of 1993 its policy toward Tajikistan subtly changed: while it still aimed to keep the Rakhmonov government in place it also pushed Rakhmonov to the negotiating table.[16]

Map 5.1 Tajikistan in Central Asia

Negotiations: Commandeering Legitimacy and Compromising the Opposition

Evgenii Primakov, at the time the Head of the Russian Foreign Intelligence Service, met with IRP chief Nuri in November 1993 in Teheran. Not coincidentally, Moscow media began broadcasting the official positions and statements of the IRP soon after the meeting.[17] Rakhmonov's hand was being forced. By March 1994 Russia's First Deputy Foreign Minister Anatolii Adimishin met with Nuri's deputy, Akbar Turadzhonzoda, and established that inter-Tajik talks would take place in Moscow under the aegis of the UN.[18]

This set in motion a series of negotiations from Moscow to Teheran to Almaty to Kabul. A three-pronged agenda emerged from the meetings, including measures aimed at a political settlement of the war, a resolution of the problem of refugees and the drafting of a new Tajik Constitution.[19] This plodding progress was secretly undermined by Rakhmonov, who suddenly declared a popular referendum for adopting a new constitution that would be coincided by popular elections for president. The opposition protested the maneuver but in the end did not have the international backing to prevent it. Despite widespread accusations of irregularities and fraud, the referendum passed and Emòmali Rakhmonov was elected president. Though admitting gross violations every international body accepted the verdict.[20] The new constitution, which was created behind closed doors without opposition participation, banned political parties based on religion. In the end, the IRP did not disengage from the negotiations, even though they never succeeded in achieving concessions from Rakhmonov on either elections or the referendum. This acquiescence will be a returning theme in IRP politics.

This trend – continuous negotiation combined with subterfuge and hypocrisy – continued all the way to 1997 when an official end to the Tajik civil war was declared. Rakhmonov was always a reluctant participant in negotiations, often granting concessions and giving ground only during those times when the United Tajik Opposition (UTO) seemed to be gaining militarily. This was the case on 23 December 1996 when he met with IRP leader Nuri in Moscow. The UTO was gaining ground in central Tajikistan and so Rakhmonov completed negotiations for protocols that set up the Commission on National Reconciliation (CNR).[21] On May 18 in Bishkek Rakhmonov and Nuri signed the Protocol on Political Questions. It was a power-sharing quota system, in many ways similar to the one Nabiev had created five years earlier and which Rakhmonov had openly mocked. In the protocol the government lifted the restrictions on UTO opposition parties on the condition that they operated within the framework of the legal system.[22]

Such wording, however, was easily manipulated by Rakhmonov. With this in mind Rakhmonov came into Moscow on June 27 and signed, along with Nuri, the General Agreement on the Establishment of Peace and National Accord. The establishment of a National Reconciliation Commission did not end hostilities, however, largely because there was a concentrated effort by the Rakhmonov government to not only undermine the peace settlement but violate it in cleverly sly ways. Thus Tajikistan 'emerged' from civil war with a compromised coalition, a co-opted Islamic party and an aggressively Machiavellian government. Understanding this foundation makes it easier to analyze the dangers intrinsic to post-civil war Tajik politics that have been unwisely ignored by American policy.

Post-Civil War Period: The Non-existent Coalition and Other Flaws

Rakhmonov was a deft politician who apparently perceived the true nature of the Islamic opposition better than the leaders of the IRP themselves. Thus from 1997 to 1999 Rakhmonov set the stage such that he was in a perfect position to utilize the 9/11 disaster to maximum benefit. The following timeline untangles the confusing web woven by Rakhmonov during this crucial post-war period, leading up to his 'free and fair re-election' in November 1999.[23]

- 1 August 1997 – Tajik parliament approves amnesty law for all UTO participants.
- 15 September 1997 – National Reconciliation Commission holds first session, four days after IRP chief Nuri returns to Dushanbe.
- 13 December 1997 – President Rakhmonov receives legislature's approval to sack eight deputies and bring charges against them for acting in anti-constitutional ways. Rakhmonov warns that if all were subject to investigation, 'half the seats in the hall would be empty'.[24]

- 15 January 1998 – UTO pulls out of National Reconciliation Commission, citing government's failure to hand over 30 percent of administration posts.

- 11 February 1998 – President Rakhmonov and IRP chief Nuri agree on which ministerial posts go to the UTO: Minister of Economy, Minister of Labor and Unemployment, Minister of Water Resources and Land Improvement, and head of Customs Committee.

- 14 February 1998 – Iranian newspaper quotes Turadzhonzoda as saying the word 'secular' should be stricken from Tajik Constitution.

- 25 February 1998 – Rakhmonov fires back, telling political activists in Khujand that the word 'secular' will not be removed.

- 10 March 1998 – Rakhmonov officially joins the Tajik People's Democratic Party (TPDP). This is the first time he has been affiliated with a party, fulfilling the constitutional obligation that a candidate must have the support of at least one registered party.

- 10 March 1998 – Turadzhonzoda takes up government post in charge of economic and trade relations with other CIS countries.

- 23 May 1998 – Tajik Parliament adopts law banning religious parties that receive 'ideological guidance' from other countries.

- 18 June 1998 – Rakhmonov endorses law banning religious parties.

- 30 July 1998 – Rakhmonov and Nuri discuss UTO posts in coalition government. Nominees fill the following posts: head of the Agriculture Ministry; chairman of the State Committee for Oil and Gas; chairman of the State Committee for Precious Stones; chairman of the State Committee for Emergency Situations and Civil Defense; chairman of the State Committee for Milk and Meat.[25]

- 14 August 1998 – UTO calls for national referendum to remove the wording 'secular state' from constitution, arguing it prevents some groups from participating in politics.

- 1 September 1998 – Itar-Tass reports that all UTO fighters in Tajikistan take an oath of allegiance to the republican government.

- 4 November 1998 – Renegade Tajik Colonel Khudaiberdiyev returns and invades northern towns and cities, says he is attacking because he deplores the criminality and the absence of democracy and free speech in Tajikistan.

- 10 December 1998 – Supreme Court bans National Unity Party.
- 10 February 1999 – National Reconciliation Council approves recommendation for creating bicameral parliament. This is the first recommendation made since it began work in September 1997.
- 3 August 1999 – IRP leader Nuri says the UTO has fully disarmed.
- 23 September 1999 – Tajik People's Democratic Party endorses Rakhmonov as its candidate in upcoming presidential elections. Rakhmonov is also made party chairman.
- 26 September 1999 – Referendum on constitutional changes held.
- 26 September 1999 – Congress of IRP meets in Dushanbe, selects Davlat Usmon to be party candidate in presidential elections. Turandzhonzoda argues for the IRP to back Rakhmonov, saying focus should be on parliamentary elections and not presidential.
- 2 October 1999 – Official results from referendum show 77.79 percent of Tajik voters approve the changes to the constitution.
- 8 October 1999 – All three alternative candidates running in the presidential race announce they are dropping out.
- 10 October 1999 – UTO pulls out of the election commission in protest at unfair elections.
- 12 October 1999 – Central Electoral Commission rejects requests from President Rakhmonov's rivals to extend time to collect the necessary 145,000 signatures to participate in elections.
- 18 October 1999 – UTO pulls out of the Reconciliation Commission
- 22 October 1999 – After the Collegium of the Supreme Court on Civil Affairs ruled that the IRP complaint about the refusal to register candidate Usmon was justified, the CEC registers Davlat Usmon as a candidate for president.
- 6 November 1999 – Presidential elections held.

- 7 November 1999 – CEC chief Mirzoali Boltuyev shows Rakhmonov as receiving 2,373,000 votes, or 96 percent of all votes cast. Usmon received 59,000 votes, or 2 percent. 98.9 percent of all possible votes in the republic are cast.
- 9 November 1999 – UTO rejoins the Reconciliation Commission.
- 16 November 1999 – President Rakhmonov inaugurated at the 14th session of the national parliament.
- 17 November 1999 – EU condemns Tajik presidential elections as 'not compatible with democratic principles and values'.[26] Says it hopes parliamentary elections will be better.

The Constitutional Referendum and Presidential Elections of 1999

What became fairly obvious to the UTO, and the IRP in particular, was that the opposition was operating at a disadvantage immediately after the signing of the Peace Accord largely because the government was not honoring its own stipulations. The lifting of bans on political parties, for example, was possible only after the second stage of the military protocol had been fully honored.[27] This meant that all UTO fighters had to disband, swear allegiance to the republican government and be put in the regular armed forces.[28] It was only then, more than two years after the signing of the Peace Accord, that the opposition ban was truly lifted.

In the intervening period clashes between UTO fighters and the government continued while Rakhmonov pushed down ever harder on the opposition. This was why the UTO repeatedly withdrew from both the Commission on National Reconciliation and the Central Electoral Commission. The crucial point was that *it didn't matter* whether the UTO was in these two bodies because Rakhmonov was going to undermine and betray the groups' missions regardless. Even the high-level postings of opposition figures were being ignored in the halls of power. Tight governmental control and a lack of access to the general media were the norm.[29]

It might seem a bit curious for the UTO to pull out of the NRC and the CEC a scant three weeks before the 1999 presidential elections. The stimulus for this self-defeating action was the public referendum on amending the Constitution pushed through on 26 September 1999. The adoption of the referendum, which passed with a rigged 92 percent popular approval, delivered a dual blow to the opposition: the presidential term of office was changed from five to seven years and the parliament was changed to a bicameral body.[30] So not only had the lead-up to the election quickly become a foregone conclusion to the opposition, but Rakhmonov at the last minute also gave himself an extra two years with which to rule. Perhaps even more damaging, however, were the long-term changes made to the legislative branch.

In this new bicameral parliament 25 percent of the deputies would be appointed directly by the president. The remaining 75 percent would be elected by indirect vote through local parliaments.[31] At face value this may not seem so bad, until it is revealed that in Tajikistan local-level parliaments were all filled by *presidential appointment*. In other words, less than six weeks before the presidential election, Rakhmonov arbitrarily changed the Tajik Constitution so that he had two more years in power as executive head *and* transformed the legislative branch so that 100 percent of its members would be de facto presidential appointees. Even though the referendum was internationally recognized as being marked by widespread proxy voting, overt falsification of voter registration lists and numerous other procedural flaws, all international bodies (including the US, OSCE and the UN) accepted the results.[32]

This is what caused the so-called 'irrational' act of pulling out by the UTO three weeks before the elections. From 1997 the international community lauded the secular-Islamic coalition as living proof of the possibility of peaceful negotiation. In reality, however, the UTO came to realize they were participating in a sham coalition that left them only futile protests and symbolic pull-outs that in the end changed nothing: Rakhmonov won the presidential election with 96 percent voter turnout and supposedly received over 96 percent of the votes cast. His closest rival received just over 2 percent. No one considered the elections free or fair yet the international community again accepted the results.

Freedom of Expression and the Media in the Post-accord Period

Another aspect of Tajikistani society that continued to garner international acclaim was the freedom Tajik media had in relation to other Central Asian states. To this day the State Department of the United States considers Tajikistan to have by far the freest of all media in the region. Perhaps the principle of relativity is important here, remembering with whom the international community is comparing Tajikistan. In closer examination it is not apparent where this reputation for freedom of expression comes from as reality does not conform to the podium declarations coming out of State.

Numbers in this case are deceiving: in August 1999 the Ministry of Culture had registered 255 publications, including 199 newspapers, of which only four were ostensibly 'controlled' by the government.[33] The problem with these impressive statistics was the fact that there was a tremendous difference between the number of newspapers formally registered with the government and the number actually *printing* papers. The Ministry of Culture itself put the figure at only 30 percent, which probably meant the number was actually closer to 15. Thus media freedom existed in a strange limbo, where the physical numbers expanded from the signing of the Peace Accord but in real-terms no one was able to produce even a single daily newspaper. Less than five published on a schedule even remotely considered regular.

Hovering over this was a government that made sure all printing presses remained state-owned and operated. Brilliant in its Machiavellian slyness, it left the government relatively free to *register* media outlets because it knew in the end the *frequency of publication* could be controlled. For whatever reason, this reality has been ignored by the international diplomatic community. For those newspapers that were registered and able to print with some semblance of regularity, the government did not leave them alone: the state continued to exert both direct and indirect pressure. Direct pressure came in one of two forms: straight-up physical intimidation or the more subtle but equally effective libel/slander/defamation of character lawsuits brought before a politically-dependent judiciary system. Rakhmonov changed the punishment for such suits so they carried jail terms up to as many as five years.[34] Indirect pressure came in one of three forms: pre-publication 'self-censorship', informal 'counseling' from governmental officials and the untimely processing of licenses and accreditation.

Newspapers and television engaged in self-censorship so as to avoid issues that put them in front of a corrupt judge. Thus 'sensitive' issues pertaining to the executive administration often got shelved in favor of less controversial subject matter. While Tajikistan in formal constitutional-legal terms appeared to indeed be one of the most advanced states in the region in regards to media freedom, these freedoms were just empty words on paper.[35] Its own constitution safeguarded the freedom of expression and forbid state censorship yet President Rakhmonov rarely acted faithfully in accordance with the constitution in such matters.

Media 'counseling' came in various forms. Sometimes Rakhmonov himself publicly admonished the media and lamented the lack of 'positive' news. Other times journalists received phone calls from the Ministry of Foreign Affairs or Justice Ministry and were advised of the danger particular stories 'represented to national security and domestic peace'. Indeed, the government was extremely quick to play the 'civil war' card to justify its own authoritarian measures. The following example of journalist Maksudjon Husseinov being summoned to the Ministry of Justice provides vivid evidence of how effective 'counseling' was:

> They asked me what I had told the Glasnost Defense Foundation in Moscow and demanded that I give them a copy of anything I had said or distributed. They also asked me why I had gone to Radio Liberty and the OSCE after my 'mugging' in October 1998. [They felt] I should have come to them. Finally they told me that were I to distribute information on media violations to human rights organizations a criminal case would be opened against me.[36]

Such counseling leaves only a rumor trail at best. Husseinov was not arrested, beaten or physically harmed. Simply the impending threat hung over his head as he left the Justice Ministry. A threat he knew the government would make good on.

The process for acquiring a license for independent television stations became increasingly labyrinthine and corrupt in Tajikistan. Both time-consuming and expensive, it required exorbitant fees and numerous bribes.[37] More remarkable was that this was actually an *improvement* over the initial procedure signed with the Peace Accord. In July 1997 the Ministry of Culture temporarily closed down most of the independent television stations for lacking a proper operating license. The problem was that at the time no government body had been established to facilitate the procedure to obtain a license. In other words, the government was shutting down free media because it failed to obtain non-existent documentation from a yet-to-be-established governing body.

The government was equally aggressive with foreign media. Its most potent weapon in this area was the revoking of accreditation. Under Tajik law journalists who visit Tajiksitan have to receive accreditation from the Ministry of Foreign Affairs.[38] What should be mere a formality often led to reporter intimidation. This was best illustrated by the case of Yelena Masyuk, a veteran war correspondent for Russia's NTV, as told by Odiljon Ashurov, a journalist in NTV's Dushanbe office:

> On 27 July 1998 Yelena Masyuk was declared persona non grata in Tajikistan for having broadcast reports discrediting the country's leadership and its policies. The government revoked her accreditation after NTV broadcast two of Masyuk's reports, on July 22 and July 23. Masyuk filed the reports from the Karategin Valley, just after the murder of four members of the United Nations Mission of Observers to Tajikistan (UNMOT) by unidentified assailants.
>
> Yelena committed an error in that she described Kulyab as a criminal center, which was a huge insult to the president: it's his home base. And she depicted the peace process as one that was hardly moving forward at all. The day after Yelena's reports were broadcast I got a phone call from the authorities. They said that they were going to close the office.
>
> Three days after Masyuk was stripped of her accreditation, NTV Dushanbe issued an apology to the government, but maintained that all facts contained in Masyuk's reports were accurate. The Ministry of Foreign Affairs also warned NTV that should they whip up of the so-called Masyuk case the Ministry [would] suspend or close the representative office in Tajikistan.[39]

Just as with domestic licensing, accreditation didn't leave any marks but surely inflicted psychological scars that negatively impacted the environment in which the media worked. Thus the immediate post-war peace in Tajikistan was marked by continued mistrust and outright degradation of the Accord protocol, even though violence never erupted. These facts were papered over by the international community. Most disturbing of all, this diplomatic apathy gave Rakhmonov the clear impression that the international community was as weak-willed as his opposition. As we shall see, this advantage did not go ignored in Tajikistan.

- 6 February 2000 – IRP leader Nuri goes to village of Buston in Sogd Province (Ferghana Valley), where his body guards and local militia are involved in a shoot-out. Two militia officers are killed.

- 7 February 2000 – Turadzhonzoda's convoy is fired upon traveling the Dushanbe-Kofarnihon road (near the Ferghana Valley region).

- 22 February 2000 – IRP leader Nuri complains that IRP representatives are not included in the local CECs in 25 single-mandate districts.

- 23 February 2000 – Turadzhonzoda warns IRP 'not to identify their narrow partisan interests with the holy faith of Islam' in the parliamentary elections.[40]

- 28 February 2000 – IRP, Communists, and Democratic Party complain about unfair elections. The CEC claims it hasn't received any formal complaints.

- 1 March 2000 – Elections held. Itar-Tass reports that the People's Democratic Party of Tajikistan (Rakhmonov's party) took 64.51 percent of the vote. The Communist party took 20.63 percent. The IRP took 7.48 percent. Out of 2,771,528 eligible voters, supposedly 2,583,909 cast ballots, an incredible 93 percent.[41]

- 6 March 2000 – Rakhmonov dismisses UTO member Davlat Usmon as Minister of Economics and Foreign Economic Relations.

- 9 March 2000 – Nomination of candidates to the 33-seat Majlisi Milli (upper parliament) begins.

- 12 March 2000 – Out of 22 seats in the Majlisi Namoyandagon (lower parliament), 15 went to the PDPT, 5 to the Communists and 2 to the IRP.

- 23 March 2000 – Elections to the Majlisi Milli are held. Only 25 seats are up for election.

- 26 March 2000 – National Reconciliation Commission disbands, saying it has fulfilled its duties.

- 29 December 2000 – Justice Ministry orders the Adolat party suspended for flagrant violations of the party's own platform and the laws of Tajikistan. Says party documents do not match reality.

A Bad Peace is Better than a Just War

The passing of the constitutional referendum, followed by Rakhmonov's victory in the presidential election and his party's dominance in parliamentary elections, completed a six-month cycle in which the president seemed intent on maintaining the appearance of democratic reform for the international community while purposefully betraying it at home.[42] The need to put up the charade was largely because Rakhmonov's regime was dependent on foreign aid.[43] Much to his delight, Rakhmonov soon learned that the western donor community, and the American foreign policy community for that matter, was easily and superficially appeased.

The March 23 (2000) parliamentary elections completed the final stipulations of the 1997 Peace Accord. It was therefore ironic that those elections also signaled the final debunking of the myth of a united Tajik opposition. While the diverse groups making up the UTO made it difficult for any coherent message and truly unified front to emerge against the president, it should not be underestimated just how much Rakhmonov himself actively pursued a strategy of divide-and-conquer, hoping to fracture and destroy any actual potential that existed.[44]

Rakhmonov's biggest step in this endeavor was when he appointed Akbar Turadzhonzoda as Deputy Prime Minister to the republican government. One of the most influential spiritual leaders within Tajikistan, Turadzhonzoda had risen to almost be considered a co-equal with Nuri in the IRP. Soon after his appointment by Rakhmonov, however, Turadzhonzoda began to openly support most of the president's initiatives and inexplicably made numerous statements to the supporters of the IRP to deemphasize Islam within its political agenda. It was Turodzhonzoda who persuaded the UTO to not boycott the 1999 referendum and presidential elections so that peace would be maintained. It was soon after these events that Turodzhonzoda was expelled from the UTO.[45]

What Rakhmonov ultimately did was play on the universal desire of Tajiks to maintain peace. Time and again, when faced by obvious subterfuge and underhanded dealing, the opposition seemed to conclude that a bad peace was better than a just war. Only Rakhmonov saw and utilized fully the potential in such war fatigue. In the four short years between the signing of the General Agreement on Peace and National Accord and the tragedy of 9/11, Rakhmonov succeeded in legally subordinating all three branches to his personal power. What's more, he did this in such a manner that the international community still lauded him for maintaining peace and 'including an openly Islamic party' in the governing coalition.

But all of this was miniscule compared to the tragedy of 9/11 and how it opened up a new arena of political opportunity that even Rakhmonov hadn't dreamed of. Unfortunately, it seems 9/11 also made the international diplomatic community even less observant over Tajik democracy than before the Twin Towers collapse. Tajik democracy has clearly paid the price as a consequence. The following timeline shows a return to assassination, arbitrary arrest and the de facto removal of the IRP from even symbolic coalition governance.

- 11 April 2001 – Deputy Interior Minister Khabib Sanginov is shot dead in his car. Sanginov was a UTO supporter who got his position as part of the peace deal.

- 24 May 2001 – Sobirjon Begjanov, a former UTO member and head of the Khujand region (Ferghana Valley), is killed in his home. He also received his position as part of the 1997 peace deal.

- 26 May 2001 – A new political party, Taraqqiyot, holds a constituent congress in Dushanbe.

- 4 July 2001 – The Ministry of Justice requests the Supreme Court's permission to ban the Adolatkhoh Party.

- 17 July 2001 – Presidential Advisor on Foreign Affairs Karim Yuldashev is assassinated in Dushanbe.

- 6 August 2001 – Supreme Court approves Justice Ministry's request to ban the activities of the Adolatkhoh party.

- 8 September 2001 – Minister of Culture Abdurakhim Rakhimov is assassinated in Dushanbe.

- 8 October 2001 – Rakhmonov confirms that the United States can use the country's airspace and some airbases for Afghan operations.[46]

- 20 February 2002 – In Brussels, Tajik Ambassador Sharif Rakhimov signs an agreement for Tajikistan to join NATO's Partnership for Peace program.

- 2 August 2002 – Chairman of the State Committee on Religion Said Akhmedov says nine imams in Isfara (Ferghana Valley) were ordered to stop conducting services because they belonged to the IRP.

- 10 January 2003 – DPT leader Makhmadruzi Iskandarov is arrested when his bodyguards are found to be carrying weapons.

- 21 February 2003 – Majlisi Namoyandagon debates lifting the clause in the constitution *limiting a president to one seven-year term.*

- 13 March 2003 – President Rakhmonov announces shadow capital amnesty from April 1 to June 1.

- 3 June 2003 – Tajik authorities announce the arrest of IRP deputy chairman Shamsuddin Shamsuddinov for 'very serious crimes'. The IRP says Shamsuddinov has been *kidnapped by the government*.[47]

- 13 June 2003 – Itar-Tass reports more than 16,000 people took advantage of the capital amnesty decree which resulted in more than 154.6 million dollars, 71.7 million rubles and 55 million sonomi returning to Tajik banks.

- 22 June 2003 – Referendum held *on 122 changes* to the constitution, including Article 65 that allows a president two seven-year terms in office. Turnout is declared at 96.39 percent with 93.13 percent voting for the changes. Only 6.13 percent voted against.

- 23 June 2003 – DPT leader Iskandarov says referendum results are false and that no more than 20 percent actually voted.

- 24 June 2003 – The OSCE releases statement saying 'the unusually high turnout of 96 percent raises concern regarding the accuracy of the reporting and results'. The statement also, however, **welcomes the calm and quiet atmosphere of the referendum.**

- 8 October 2003 – Prosecutor-General's office hands over criminal case against IRP deputy leader Shamsuddinov to the Supreme Court, accusing him of organizing an illegal armed group, treason, crossing the border illegally and bigamy.

Intensifying the Repression against Islamists (and everyone else for that matter)

It was clearly an objective of Rakhmonov to attack the Peace Accord in stages. During the first stage (1997–2000) it was necessary to placate the international community while preventing the opposition from gaining any real traction. This was most effectively done by isolating UTO postings while passing the referendum that extended his presidency and made the legislature an executive rubber stamp.

During the second stage (2001–2003) Rakhmonov maximized his leverage. Many UTO officials were summarily removed while yet another constitutional referendum removed presidential term-limits, making Rakhmonov eligible to serve *another* two terms. Tajikistan is thus likely to end up with Rakhmonov holding the presidency for a generation – a total of 26 years.

What was also obvious was the continued harassment and in some cases assassination of UTO/IRP deputies. While impossible to definitively prove executive culpability, it is clear the environment created by Rakhmonov did not stabilize Tajikistani politics but actually made them more volatile, at least for the opposition. This 'wearing down' had clearly made its mark by 2003: almost all opposition figures in governing positions had become *proponents* for most of the president's policies and were consistently in conflict with *their own opposition parties*.

Most importantly disturbing for the present work was the noticeable apathy and acquiescence by the international community. It continually accepted the results of referendums and elections that were considered grossly falsified and unfair by their own international observers. This acceptance continued in spite of the fact that Rakhmonov's maneuvers became far less subtle after 9/11. Most dissatisfying of all was the fact that no diplomatic circle was more supportive and laudatory than the American contingent.

It was Rakhmonov who, emboldened by the international war against terrorism and the sudden support he was receiving from the US, declared that the IRP had 'some members who, contrary to Tajikistan's law on political parties, were indoctrinating people in the spirit of extremism'.[48] Shortly thereafter, 'proficiency tests' for Islamic spiritual leaders began being administered throughout the republic to the heads of over 250 mosques and 20 religious schools. The aim of the tests was to ensure these religious leaders had an advanced knowledge of Tajik secular law. In addition, all those forced to take the test had to swear an oath of allegiance to the current administration. Simultaneously, Rakhmonov attacked the northern stronghold of the IRP, closing down numerous mosques because there were supposedly 'too many in the area'.[49] None of this seemed to bother the international community, which as a consequence looked complicit in the repression. Instead of being a bastion of democratic development and a model for incorporating Islamic parties into the political framework (as international rhetoric constantly proclaimed), Ferghana radicals saw the West as tolerant and even supportive of a Tajik tyrannical government.

The lesson learned by Rakhmonov was obvious: why strive to achieve a real secular-Islamic coalition when the West was so eager to praise and provide support to a fake one? The current stage (2003–present) unfortunately shows no end to the decaying trends.

- 13 January 2004 – Supreme Court sentences IRP deputy leader Shamsuddinov to 16 years imprisonment.[50]

- 15 January 2004 – IRP complains about the court decision against Shamsuddinov and sends protest to the UN and other guarantors of the Peace Accord. Nothing comes from international bodies.

- 19 January 2004 – President Rakhmonov dismisses Nigina Sharipova as deputy prime minister. Names Abduljabor Rakhmonov as new head of state television and radio broadcasting.

- 26 January 2004 – President Rakhmonov transforms the Presidential Guard into the National Guard, dismisses Gafor Mirzoyev and names Colonel Rajabali Rakhmonaliyev to take his place.

- 15 March 2004 – EU External Relations Commissioner Chris Patten says in Almaty that Central Asian governments need to respect human rights and democracy more, as these are ‘precisely the conditions which breed and nurture the hate and grievance that underlie terrorism’.[51]

- 29 March 2004 – Prosecutor-General Bobokhonov gives press conference to reveal the arrest of twenty members of the Islamic organization Bayat. All have been charged with the murder of a Baptist pastor and setting fire to mosques where imams loyal to the government were preaching.

- 12 June 2004 – Leaders of numerous opposition groups file an appeal to President Rakhmonov, expressing concern about the adoption of a new election law. None of the opposition’s provisions made it into the final draft, including the presence of registered political parties in all CECs, that ballots be protected, that the use of pencils for filling in ballots be prohibited and that copies of approved ballot-counting protocols be given to election observers. Rakhmonov ignores them.

- 16 June 2004 – The Majlisi Namoyandagon, the lower house of parliament, adopts the election law without any opposition provisions. The law introduces ‘fees’ for registering as a candidate, including a general 300 USD fee for any candidate to run.[52] This is regarded as a compromise as the PDPT, Rakhmonov’s party, wanted to institute a 3,000 USD individual fee and an 80,000 USD party fee.

- 15 July 2004 – President Rakhmonov signs the bill into law.

- 20 July 2004 – The IRP criticizes the new law, saying the ‘lesser’ registration fee is equal to 200 minimum wages and in general ‘does not allow 80 percent of the population to be elected’.[53]

- 18 August 2004 – The National Association of Independent Mass Media Outlets of Tajikistan (NANSMIT) releases a statement saying problems with access to information are becoming more frequent as parliamentary elections in 2005 approach.

- 27 August 2004 – Khusein Okilov, a senior official in the Tax Ministry, says the printing office working for four major opposition papers was targeted and shut down by the government for tax reasons, not political ones.

- 19 November 2004 – Authorities ban distribution of main opposition newspaper 'Ruzi Nau'.[54]

- 9 December 2004 – Leader of the Democratic Party of Tajikistan (DPT) Makhmudruzi Iskanderov detained in Moscow on charges of committing terrorist acts and illegal transport of weapons.

- 18 December 2004 – DPT holds congress, call on the United States, OSCE, UN and EU to help free Iskanderov, calling his arrest politically motivated. Statement says 'if the *international community does not defend democracy in Tajikistan and does not take measures to free Iskanderov then this will mean the recognition of totalitarianism in Tajikistan'*.[55]

- 14 January 2005 – CEC chairman Mirzoali Boltuyev confirms DPT leader Iskanderov will not be registered as a candidate for parliamentary elections due to the criminal charges.

- 26 January 2005 – IRP says only 8 of 23 IRP candidates have succeeded in getting registered under the new election law.

- 9 February 2005 – Chairman of the Social-Democratic Party Rakhmatullo Zoirov says state media and government have been gradually connecting the current political opposition to the former 'opposition' of the civil war period of 1992–1997.

- 27 February 2005 – Elections to the Majlisi Manoyandagon held. Out of 63 total seats, the PDPT, Rakhmonov's party, ultimately takes 52 of them by the time the second round is held in March.

- 5 October 2005 – Celebrations and events are planned by the government as 2006 is declared the 'Year of Aryan Civilization'. Islamic groups protest to no avail what it claims is a cynical political manipulation in order to provide alternative religious distraction away from Islam in the lead-up to the 2006 presidential elections.

- 28 October 2005 – The IRP decries recent Ministry of Education initiatives that first ban female students from attending educational institutions while wearing a head scarf and prohibits *all children under the age of 16* from attending mosques.

- 20 January 2006 – British Embassy in Dushanbe formally protests what it considers to be an illegal and politically motivated shut-down of BBC radio broadcasts across Tajikistan.

Bread, Circuses and Confrontation

By the time the 2005 parliamentary elections rolled around Rakhmonov had long since decimated the presence of the UTO in general and the IRP in particular. No less significant, however, was the fact that even old *allies* were finding themselves removed even though the peace process had allowed many warlords to effectively retain control over numerous regions, including the influential Kulyab and Pamir regions.[56]

Rakhmonov's strategy was to start slowly, subtly, moving the less vocal and less visible figures out of the command structures first. Then, with momentum building, he was able to concentrate on more significant and powerful figures. When it came to the IRP this usually meant they were removed from office on charges of 'incompetence' or, worse still, arrested on charges of treason and/or corruption. In the world of post-Soviet politics, these reasons were the easiest ones to conjure out of thin air. For the warlords who helped Rakhmonov initially come to power, they were arrested, imprisoned and, in some cases, conveniently murdered, always with their various 'criminal associations' taking the blame.[57]

While it is not the goal to drum up conspiracy theories involving President Rakhmonov and murder, it is nonetheless curiously fascinating that so many people were assassinated *after* the National Reconciliation Commission folded its tent and declared all of the protocols of the Peace Accord fulfilled. Just as Tajikistani politics were meant to be stabilized and heading towards consolidation, Rakhmonov intensified his efforts at confrontation.

Rakhmonov's consolidation of power revolved around his home town of Danghara. A perfect example of this was the removal of Mirzoyev as head of the Presidential Guard. When Rakhmonov transformed it into the National Guard he put Rakhmonaliyev, a native from Danghara, in charge.[58] IRP leader Nuri made an accurate and pointed perception about this shift to 'new' officials wholly dependent on the president:

> Those who made peace, from both the government and the opposition, always consider peace to be the most important thing of all. But others have appeared in law enforcement [and elsewhere] since then and most of them had no role in bringing about peace. They don't understand the meaning of peace or what can destroy it.[59]

The IRP's position in terms of power was compromised at every turn. Rakhmonov's government never missed an opportunity to capitalize on the 'civil war' card or 'extremism' card, with the IRP apparently too weak in political will to counter it. When it was revealed three detainees at Camp X-ray in Guantanamo Bay, Cuba were actually Tajiks from the Ferghana Valley stronghold of Isfara, Rakhmonov denounced local officials for their permissiveness in allowing radical groups to proliferate.[60] These comments marked the beginning of a religious crackdown that only served to further radicalize the hidden groups already fomenting within the Ferghana. As a result, the intensity and attractiveness of radical groups have only increased since 9/11 in Tajikistan.

Rakhmonov also employed the 'bread and circuses' strategy after 9/11. In an obvious attempt to placate and distract the local population, the government announced in 2002 that 2005 would be the 2,700th anniversary of Kulob, the home city of Rakhmonov's native clan. Massive construction projects were announced with many apartment buildings and storefronts renovated. Teahouses and western-style supermarkets were built in the name of the coming celebration. But all of this was mere window-dressing that did not easily fool locals:

> They'll build a supermarket that nobody will ever use. They'll build a teahouse that will be empty within a year. Kulob already has teahouses and nobody goes to them anyway. Why will they be any different? Wouldn't it be better for them to build a factory or two so that people will have jobs? Or renovate some schools? This would win them the love of the people of Kulob.[61]

Rakhmonov and the Tajik government did not *care* to win the love of the people of Kulob. It was not necessary to secure power. The symbolic manifestation of improvement was more relevant and strategic. As Iskanderov, the DPT leader first arrested in Moscow and then *kidnapped* by Tajik authorities, declared in 2004:

> There is no difference between the PDPT and the old CPSU (Communist Party of the Soviet Union). If you want any kind of position, you absolutely have to join the president's party. Regional governors, mayors, ministers – anybody who wants to have a career has to join the PDPT.[62]

This assessment was confirmed with the lead up to and results of the 2005 parliamentary elections. Not only did Rakhmonov's PDPT take 52 out of 63 possible seats, once again registration procedures were tinkered with before the elections, making it more difficult for opposition parties to be legally registered. This was the case with the Taraqqiyot party which tried unsuccessfully for *four years* to register. After various delay tactics were on the verge of running out the Justice Ministry came forward with the announcement that *eleven* members on a party list of nearly 2,000 denied being actual members.[63] On account of this 'discrepancy' the party was ostensibly banned from the political process.

Such events signal to the opposition that symbolic pluralism in Tajikistan was acceptable as long as it did not challenge the status quo in any real way.[64] What did not bode well for Tajikistan democracy was that these unwritten repressive rules had long been acquiesced to by the 'legal' Islamic opposition, the IRP. It is not difficult to conclude such acceptance only pushes other groups further from the mainstream, away from the principles of negotiation and compromise and even away from the concept of moderate Islam. And this quite frankly was understandable because in Rakhmonov's world the only thing to negotiate was how much domination you would accept and the only thing to be compromised was the opposition:

> Rakhmonov doesn't seem to know what's going on in Tajikistan…The Nobel Prize nomination,[65] all these [fake] government holidays – he has no idea about reality. The government papers print that all is well, the TV programs all praise him… I think this is probably from the people around him – they'll do anything not to spoil his mood.[66]

The above criticism is confirmed by Rakhmonov himself when he spoke about the Kyrgyz regime change that took place in March 2005:

> We can say with confidence that Tajikistan is moving along the right path. The development of a mechanism of collaboration between the executive and legislative branches ensures the stepping up of political and economic reform.[67]

This comes from a man who had already subjugated the legislative branch to his executive power, had begun pressuring Tajik democracy advocates to not interact with international organizations and had arrested potential candidates seeking to challenge him in the 2006 presidential elections.[68] It was not surprising that a tyrant did not notice his own transgressions. It was surprising, however, that the United States did not seem to notice, or, worse still, did not seem to truly care. What conclusions can be drawn by comparing developments in Tajik politics with the evolution of US policy toward the state since 9/11? Unfortunately, the trend discovered and already detailed in Bishkek is even worse in Dushanbe.

US Policy: Talking the Talk but Not Walking the Walk

There was an ominous change in US policy after 9/11 when the US lifted an eight-year old arms sales ban to Tajikistan in January 2002.[69] This was done as a reward for its support in the war against terror. The problem was that this was already well into Rakhmonov's consolidation strategy. The United States was agreeing to sell weapons to a state that had in the previous five years completely undermined its democratic foundations.

Joost R. Hilterman, Executive Director of the Arms Divison of Human Rights Watch accurately criticized the move, saying, 'these transfers won't make the United States more secure in the long run…and they make the United States *complicit in the abuse of civilians in other countries'*.[70] In addition, the US set up mechanisms to 'fast-track' military assistance so as to provide aid that would normally take more than a year to expedite.[71] Disturbingly, the only check on these deliveries seemed to be a State Department 'urging' that the recipients respect human rights. Other than that, the US made little effort to ensure its weapons would actually be used judiciously and exclusively for anti-terror and crime prevention. Considering the maneuvers that took place in Tajikistan before and after the lifting of the ban, it seems this blind faith was horribly misplaced.

US policy towards Tajikistan was testimony to 'Wonka Vision', the only uniqueness being the added decorative mantra that praised the secular-Islamic coalition. The Bureau of European and Eurasian Affairs gave tribute to both Tajikistan and American policy in early 2003, despite the fact that nothing it discussed seemed to conform to reality:

> Improvements in security allowed the [Tajik] government to continue its halting *progress towards further democratization and rule of law* (italics mine). The government issued the first license for an independent radio station in the capital…The government amended the national law on media but it did retain *some flawed portions of the law*. Though new opposition parties still experienced difficulties in registering, *Tajikistan is still the only country in Central Asia with an openly religious opposition party* – the Islamic Renaissance Party of Tajikistan – which holds seats in the legislature.[72]

From a policy standpoint it is clear the United States is willing to discard empirical facts in order to 'stay on the talking points'. For Tajikistan this meant two things: first, emphasize the halting progress and second, hammer home repeatedly that Tajikistan was the only country with a legal Islamic party. Of course, that legal representation consisted only of *two* seats in a compromised parliament.

US Wonka Vision runs rampant in State Department memos and media communiqués, constantly referring to the 'open' Tajik political environment, the 'free media' and the 'vibrant political space' the opposition has carved out for itself.[73] In short, these are fantasies. There is no free media in Tajikistan. It has no open political environment and the opposition has lost its political space. And yet still the US drones on with its coalition mantra because the higher priority is maintaining convenient geographical allies in the war against terror, regardless of the long-term consequences to said war.

In 2003 and 2004 State Department assessments admitted to some of the 'secretive' and 'unfortunate' initiatives transpiring within the Rakhmonov regime, but the criticism was always muted and the transgressions compensated by other supposed signs of progress. In the end, it was most important that 'Tajikistan was a coalition member of the Global War on Terrorism and had fully supported US security objectives in the region' and that the 'political will and capability on the part of the Tajik government to assist in the Global War on Terror' had increased *because of USG assistance*.[74] Our short-term priorities trumped other considerations.

Keep in mind this is not a plea for pacifism. I focus instead on the fact that our policy toward Central Asia, and Tajikistan in particular, is *an actual violation of American law that can ultimately undermine victory in the war against terror*.

> Section 201 of the Freedom Support Act amended Section 489A of the Foreign Assistance Act of 1961 to require the President to take into account not only relative need but also the extent to which that independent state is acting to make significant progress toward and is committed to the comprehensive implementation of a democratic system based on the principles of the rule of law, individual freedoms, and representative government.[75]

The main financial mechanism to funnel aid to Tajikistan has been the Freedom Support Act. Even before 9/11 Tajikistan did not meet the requirements of Section 201. After 9/11 the trends only worsened and yet US military and financial assistance tripled. The official explanation supposedly satisfying the requirements of Section 201 read as follows:

> The government established some democratic institutions. While key elements of the central government continued to be directly or indirectly controlled by President Rakhmonov, some politicians from opposing parties continued to hold seats in the parliament and positions in the government. The government and wartime opposition continued to cooperate on key issues…It is noteworthy that Tajikistan has a legal opposition, including a political party that is both overtly Islamic and committed to a secular state.[76]

This is Wonka Vision: official government policy simply sees what it wishes to see. The State Department even used spin-control to praise the cessation of assassination in Tajikistan even though it ignored the fact that the outbreak of assassinations came a full *four years* after the signing of the Peace Accord and well into the so-called stabilization period.[77]

The Freedom Support Act itself is a partial enabler by providing contradictory provisions. For example, the criteria quoted above was Section 489A(1). Section 489A(8) deals with the 'denial of support for acts of international terrorism'. On this level, Tajikistan has proven itself quite valuable.

> The government of Tajikistan does not grant sanctuary from prosecution to individuals or groups that have committed acts of international terrorism or otherwise support international terrorism. Tajikistan was strongly critical of the former Taliban in Afghanistan for harboring Osama bin Laden. It did not negotiate with the Taliban. After the attacks of September 11 Tajikistan immediately supported Coalition forces in Operation Enduring Freedom and has generously offered assistance without reservation.[78]

Even worse is when Section 489A(b)(1) is read in combination with the official Bush administration justification:

> Has the President determined that the government of Tajikistan has engaged in a consistent pattern of gross violations of internationally recognized human rights or of international law?
>
> No. The President has not made such a determination at this time. While there have been serious shortcomings in human rights observance in Tajikistan, the government has undertaken effort to address some of the problems…We will work to better address these human rights problems not only through diplomatic efforts but also through our assistance programs.[79]

The problem of course is that Tajikistan has engaged in a gross pattern of rights violations, shortcomings have not been addressed and the US has not better addressed the issues over time through diplomatic efforts and assistance programs. From the signing of the Peace Accord through to the present day, Tajikistan democracy has decreased. It only worsened after the start of the 'Global War on Terror' and the initiation of intensified US engagement. Under such discrepancies US policy is clear for all to see: you only need to talk the democratic talk, but you must walk the anti-terror walk.

Such a policy is dangerous regardless of country. In Tajikistan, however, it is doubly dangerous because of the so-called 'secular-Islamic coalition'. A careful analysis of political Islam in Tajikistan reveals such co-opted weakness that the IRP itself may be unwittingly contributing to Islamic fundamentalism.

The Intrinsic Impotence of Political Islam in Tajikistan

The foundation for many of the problems inherent to contemporary Islam within Tajikistan was actually laid during Soviet times. There was always a dual attitude towards Islam, one epitomized by a fierce struggle against and the other involving tacit acceptance of a softened version of the religion.[80] Soviet authorities aimed at the destruction of institutional clergy because this was thought to be the best way to eliminate the desire for Muslim law within Tajikistan. What came to be known as 'popular Islam' was allowed to exist but only through a conformist clergy that had already been thoroughly vetted by Soviet authorities.[81]

This destruction was not total, however. Repressing the institutionalized clergy and placing strict Soviet control over religion only served to drive many true believers deep into hiding. This was the delicate balance always present in Tajikistan and it remains today: the overt clergy are not necessarily regarded as the true religious authority while an underground presence of radicalized Islamists has carved out a secret niche deep within society.

The elite clan structure common across all of Central Asia was certainly a backbone for the formation of the Tajik nomenclature. While the northerners fought for greater autonomy from the very beginning of the state, they also never actively sought to make a clean break from Moscow. Thus in the regions where the Communist Party came to dominate – the Gissar Valley, the Leninabad region and in the Kulyab – people came to be closely bound with both the state and this 'vanilla' form of political Islam.[82] These regional groups, who would form the base of Tajikistan's post-Soviet elite, supposedly always had a greater feeling of harmony between secular and religious power, with the former always maintaining a higher position over the latter.

In the past this has been taken as a great positive in Tajikistani politics. I argue in fact that this harmony was and remains simply the subordination of the religious to the secular. It is not a harmonization at all. Harmony 'gets along' because the Islamic side of the equation has been mutated into a much lesser version of itself. Thus, when the IRP became a 'significant force' in Tajikistani politics after the

1997 Peace Accord, its significance did not come because of its religiosity but *in spite* of it. It was its willingness to rigidly be a political party first and a religious party second that spared it the most severe rods of Rakhmonov's repression. It is not coincidental that experts in the region feel the strongest clash in the country is not between secular forces and Islamic ones but between 'popular' Islam and more fundamentalist interpretations.[83] In Tajikistan, as painful as it may be for the West to admit, the fundamentalist version of Islam comes off as much more legitimate. Our engagement with the country, whether we choose to admit it or not, has only exacerbated this tension and given the fundamentalists even more legitimacy.

The Islamic Renaissance Party of Tajikistan was registered by the Ministry of Justice on 4 December 1991. The main objectives of the IRP were three-fold:

- The spiritual rebirth of the citizens of the republic
- The economic and political independence of the republic
- The political and legal awakening with the purpose of implementing the fundamental provisions of Islam in the life of all Muslims.[84]

In short, its program was compromised Islam. The party has maintained since its registration that the true goal of the IRP was to 'create a democratic state based on the rule of law with a government of popular confidence comprised of representatives of various political forces'.[85] It is a difficult task to find within that chief objective the ISLAMIC part of the IRP's mission. If this is the Islamic half of the secular-Islamic coalition in Tajikistan how does the IRP represent Islam with such views? Even Muslim groups only slightly right of center in Tajikistan do not see a legitimate Islamic authority in the IRP. And in that is the chief problem: the IRP, which is still too Islamic for Rakhmonov, is not nearly Islamic enough for the general population to truly represent Islam in a secular-Islamic coalition. What is still regarded today in the international community as a momentous step in Central Asian democratic development is in reality a non-measure that fomented a permanent radical Islamic underground within Tajikistan and cemented IRP's compromised status.

When the ninth session of the Majlis Oliy parliament opened in May 1998 a law was proposed that would ban all political parties from functioning on a religious basis.[86] The IRP rightfully protested the law and succeeded in having it redrafted. In November during the tenth session a new version was passed that formally recognized the right of the IRP to exist and participate in politics. However, amendments to Article 4 of that law *banned mosques and madrasahs from setting up primary organizations of a political and military-political nature.*[87]

In other words, it is not that the IRP won the right to be the voice of Islam in Tajikistan from Islamists. It won the right from the repressive government that despises Islam. Only the IRP was accepted as Islamic. No other voices were allowed to be heard, at least not in politics. The secular government determines for itself what is acceptably Islamic while all others are cast out. It is this environment, and the IRP's acceptance of it, that does not bode well for the future of a true secular-Islamic coalition. The IRP cannot help but be viewed as a tacit supporter in the repression of its own religion's political freedom.

Crucial in all of this is not the leadership of the IRP but the mass followers of the party. The IRP has grown its base exactly in the areas where conservative Islam is strongest. The people originating in and around the Ferghana Valley have increasingly become one of the most important popular bases of support for the IRP.[88] But as this book points out, the Ferghana Valley's conservative Islam is not coincident with the Islam deemed acceptable by Rakhmonov and the international community. Thus it is easy to foresee a moment down the road where the ideology of the IRP, which has always maintained a formal conservative Islamism in its rhetoric, clashes with the actual political achievements of the party. This clash, when it comes, will come from the rank and file members expressing dissatisfaction and possibly even betrayal at the hands of their own leadership. This criticism already exists amongst the radical splinter groups that have formed since Rakhmonov's increased repression after 9/11. It is likely a question of when and not if the IRP will ultimately face an internal explosion for the soul of the party. Too many times the IRP has decided it was best to maintain a 'low profile' and honor the ideal of peace at any cost.[89] It is always just such decisions that inflame radicals to greater and more intense hatred. The IRP's moderation can be admired for its avoidance of bloodshed but it has not been met halfway by Rakhmonov. As a result, to radical Islamists the IRP looks like a political flunky willing to compromise its faith in return for nothing. And it does not go unnoticed by radicals how often the US has praised this very coalition.

It has been a very delicate balance for the IRP, one not to be envied. Its effort to legitimately remain a part of the political process has been mocked by a president willing to do anything to maintain his own iron-grip on power. Its desire to maintain a commitment to peace through open dialogue and negotiation has earned it rave reviews from the international community. Yet this same commitment has also made the party a symbol of co-optation and prone to accusations of being a sell-out by those it would consider its rank and file foundation. In the evolution of Tajikistani politics since the Peace Accord the IRP has become more of a purely political party and not an Islamic one. It has lost along the way most of its true 'political Islamic nature'.[90] This has left it acceptable to both the local elites in power and the Western diplomatic community but it has largely isolated the party from the larger international Islamic network, which no longer sees a kindred spirit in the IRP.

Islamic parties like the IRP need to be careful of their bedfellows: US support will often wax poetic in praise but in real-terms will tend to be fickle in concrete commitments. Arab Islam, on the other hand, tends to react violently when betrayal is perceived against the faith. The IRP is close to being judged perpetrators of such betrayal. The IRP should remind itself of the fate of Ahmad Shah Massoud, the Northern Alliance leader in Afghanistan. As he became more moderate he also garnered great admiration from the West. That admiration, however, did not trigger outright support and much-needed aid.[91] Massoud was assassinated by followers of

Osama bin Laden on 9 September 2001, just two days before the Trade Center tragedy.

Conclusion – The Fog of Secular-Islamic Coalition

One must not forget the point behind the original structure of the 1997 Peace Accord. Part of the reason to install a quota system was to gradually institutionalize the idea of peaceful negotiation and loyal opposition. This was in fact never possible in Tajikistan because Rakhmonov never intended to see such institutionalization take place. The accord was a charade meant to dupe not only the international community but also to lull the opposition into complacency. By the time the latter was more aware of the systematic effort being perpetrated by Rakhmonov it was already too late: he altered the composition and allegiance of the legislative branch, extended his term to seven years and made himself eligible for two more terms, eliminated the opposition from truly participating in the coalition and then de facto removed them from government altogether.

All of this makes logical sense in a perfect Machiavellian world of absolute power. But it is highly dangerous in a post-9/11 world where the US looks for allies in the war against terror and finds people like Rakhmonov waiting with open arms. Up to this point, despite empty rhetoric voicing 'concern' and 'puzzlement', we have accepted his embrace. It would be naïve to think radical groups in the area have not noticed. What this chapter has attested to and what careful analysis reveals is an unholy trinity of factors in Tajikistan that all work to radicalize Islamists – the impotence of the IRP, brazen repression from Rakhmonov and relative tolerance by US politicos – all come together to create a volatile cauldron of hate aimed not just at Tajikistan's government but at the US as well. The fog of our praise for Tajikistan's secular-Islamic coalition has only succeeded in making us unable to see the true dangers lurking within.

---

1 Arkadii Dubnov, 'Tajikistan', The Rand Corporation, 1996.

2 Arkadii Dubnov, 'Tajikistan', The Rand Corporation, 1996.

3 Arkadii Dubnov, 'Tajikistan', The Rand Corporation, 1996.

4 Arkadii Dubnov, 'Tajikistan', The Rand Corporation, 1996.

5 Arkadii Dubnov, 'Tajikistan', The Rand Corporation, 1996.

6 Arkadii Dubnov, 'Tajikistan', The Rand Corporation, 1996.

7 Arkadii Dubnov, 'Tajikistan', The Rand Corporation, 1996.

8 Accord: An International Review of Peace Initiatives, 'A Chronology of Events in Tajikistan', http://www.c-r.org/accord/Tajik/accord10/chronology.shtml, accessed 14 December 2004.

9 Accord: An International Review of Peace Initiatives, 'A Chronology of Events in Tajikistan', http://www.c-r.org/accord/Tajik/accord10/chronology.shtml, accessed 14 December 2004.

[10] Accord: An International Review of Peace Initiatives, 'A Chronology of Events in Tajikistan', http://www.c-r.org/accord/Tajik/accord10/chronology.shtml, accessed 14 December 2004.
[11] Arkadii Dubnov, 'Tajikistan', The Rand Corporation, 1996.
[12] Arkadii Dubnov, 'Tajikistan', The Rand Corporation, 1996.
[13] Accord: An International Review of Peace Initiatives, 'A Chronology of Events in Tajikistan', http://www.c-r.org/accord/Tajik/accord10/chronology.shtml, accessed 14 December 2004.
[14] Accord: An International Review of Peace Initiatives, 'A Chronology of Events in Tajikistan', http://www.c-r.org/accord/Tajik/accord10/chronology.shtml, accessed 14 December 2004.
[15] Arkadii Dubnov, 'Tajikistan', The Rand Corporation, 1996.
[16] Arkadii Dubnov, 'Tajikistan', The Rand Corporation, 1996.
[17] Arkadii Dubnov, 'Tajikistan', The Rand Corporation, 1996.
[18] Arkadii Dubnov, 'Tajikistan', The Rand Corporation, 1996.
[19] Arkadii Dubnov, 'Tajikistan', The Rand Corporation, 1996.
[20] Accord: An International Review of Peace Initiatives, 'A Chronology of Events in Tajikistan', http://www.c-r.org/accord/Tajik/accord10/chronology.shtml, accessed 14 December 2004.
[21] Accord: An International Review of Peace Initiatives, 'A Chronology of Events in Tajikistan', http://www.c-r.org/accord/Tajik/accord10/chronology.shtml, accessed 14 December 2004.
[22] Accord: An International Review of Peace Initiatives, 'A Chronology of Events in Tajikistan', http://www.c-r.org/accord/Tajik/accord10/chronology.shtml, accessed 14 December 2004.
[23] All timeline dates and factoids are from Radio Free Europe/Radio Liberty, http://www.rferl.org, accessed from August 2004 to November 2005.
[24] Radio Free Europe/Radio Liberty, 'Tajikistan Votes 2005', http://www.rferl.org, accessed 17 November 2005.
[25] Radio Free Europe/Radio Liberty, 'Tajikistan Votes 2005', http://www.rferl.org, accessed 25 November 2005.
[26] Radio Free Europe/Radio Liberty, 'Tajikistan Votes 2005', http://www.rferl.org, accessed 03 December 2005.
[27] Human Rights Watch, 'Tajikistan Report 1999', http://www.hrw.org, accessed 23 March 2002.
[28] Human Rights Watch, 'Tajikistan Report 1999', http://www.hrw.org, accessed 23 March 2002.
[29] Human Rights Watch, 'Tajikistan Report 1999', http://www.hrw.org, accessed 23 March 2002.
[30] Human Rights Watch, 'Tajikistan Report 1999', http://www.hrw.org, accessed 16 April 2003.
[31] Human Rights Watch, 'Tajikistan Report 1999', http://www.hrw.org, accessed 16 April 2003.
[32] Human Rights Watch, 'Tajikistan Report 1999', http://www.hrw.org, accessed 16 April 2003.

[33] Human Rights Watch, 'Tajikistan Report 1999', http://www.hrw.org, accessed 08 November 2003.

[34] Human Rights Watch, 'Tajikistan Report 1999', http://www.hrw.org, accessed 08 November 2003.

[35] Human Rights Watch, 'Tajikistan Report 1999', http://www.hrw.org, accessed 08 November 2003.

[36] Human Rights Watch, 'Interview with Maksudjon Husseinov – Tajikistan Report 1999', http://www.hrw.org, accessed 14 November 2003.

[37] Human Rights Watch, 'Tajikistan Report 1999', http://www.hrw.org, accessed 20 November 2003.

[38] Human Rights Watch, 'Tajikistan Report 1999', http://www.hrw.org, accessed 21 November 2003.

[39] Human Rights Watch, 'Revocation of Accreditation', http://www.hrw.org/reports/1999/Tajikistan/Tajik99n-08.htm#TopOfPage, accessed 28 November 2004.

[40] Radio Free Europe/Radio Liberty, 'Tajik Elections', http://www.rferl.org/specials/Tajikelections/timeline2000.asp, accessed 12 January 2004.

[41] Radio Free Europe/Radio Liberty, 'Tajik Elections', http://www.rferl.org/specials/Tajikelections/timeline2000.asp, accessed 12 January 2004.

[42] Central Asia-Caucasus Analyst, 'Rakhmonov's Election Maneuvers and Tajikistan's Future', http://www.cacianalyst.org/view_article.php?articleid=200, accessed 5 May 2003.

[43] Central Asia-Caucasus Analyst, 'Rakhmonov's Election Maneuvers and Tajikistan's Future', http://www.cacianalyst.org/view_article.php?articleid=200, accessed 5 May 2003.

[44] Noor Umarov, Central Asia-Caucasus Analyst, 'Does Tajikistan's Parliamentary Elections Spell the End of the UTO?', http://www.cacianalyst.org/view_article.php?articleid=179, accessed 12 July 2004.

[45] Noor Umarov, Central Asia-Caucasus Analyst, 'Does Tajikistan's Parliamentary Elections Spell the End of the UTO?', http://www.cacianalyst.org/view_article.php?articleid=179, accessed 12 July 2004.

[46] Radio Free Europe/Radio Liberty, 'Tajik Elections', http://www.rferl.org/specials/Tajikelections/timeline2001.asp, accessed 03 January 2005.

[47] Radio Free Europe/Radio Liberty, 'Tajik Elections', http://www.rferl.org/specials/Tajikelections/timeline2001.asp, accessed 03 January 2005.

[48] Human Rights Watch, 'Tajikistan Report 2002', http://www.hrw.org/wr2k3/europe12.html, accessed 01 April 2004.

[49] Human Rights Watch, 'Tajikistan Report 2002', http://www.hrw.org/wr2k3/europe12.html, accessed 01 April 2004.

[50] Radio Free Europe/Radio Liberty, 'Tajik Elections', http://www.rferl.org/specials/Tajikelections/timeline2004.asp, accessed 09 February 2005.

[51] Radio Free Europe/Radio Liberty, 'Tajik Elections', http://www.rferl.org/specials/Tajikelections/timeline2004.asp, accessed 09 February 2005.

[52] Radio Free Europe/Radio Liberty, 'Tajik Elections', http://www.rferl.org/specials/Tajikelections/timeline2004.asp, accessed 09 February 2005.

[53] Radio Free Europe/Radio Liberty, 'Tajik Elections', http://www.rferl.org/specials/Tajikelections/timeline2004.asp, accessed 09 February 2005.

[54] Radio Free Europe/Radio Liberty, 'Tajik Elections', http://www.rferl.org/specials/Tajikelections/timeline2004.asp, accessed 09 February 2005.

[55] Radio Free Europe/Radio Liberty, 'Tajik Elections', http://www.rferl.org/specials/Tajikelections/timeline2004.asp, accessed 09 February 2005.

[56] International Crisis Group, 'Tajikistan's Politics: Confrontation or Consolidation?', Dushanbe/Brussels, 19 May 2004.

[57] International Crisis Group, 'Tajikistan's Politics: Confrontation or Consolidation?', Dushanbe/Brussels, 19 May 2004.

[58] International Crisis Group, 'Tajikistan's Politics: Confrontation or Consolidation?', Dushanbe/Brussels, 19 May 2004.

[59] International Crisis Group, 'Tajikistan's Politics: Confrontation or Consolidation?', Dushanbe/Brussels, 19 May 2004.

[60] International Crisis Group, 'Tajikistan's Politics: Confrontation or Consolidation?', Dushanbe/Brussels, 19 May 2004.

[61] International Crisis Group, 'Tajikistan's Politics: Confrontation or Consolidation?', Dushanbe/Brussels, 19 May 2004.

[62] Interview with Makhmudruzi Iskanderov, International Crisis Group, Tojikobod, Tajikistan, March 2004.

[63] International Crisis Group, 'Tajikistan's Politics: Confrontation or Consolidation?', Dushanbe/Brussels, 19 May 2004.

[64] International Crisis Group, 'Tajikistan's Politics: Confrontation or Consolidation?', Dushanbe/Brussels, 19 May 2004.

[65] This refers to an effort in 2001 to have Rakhmonov nominated for the Nobel Peace Prize.

[66] Interview with independent journalist, Dushanbe, International Crisis Group, 'Tajikistan's Politics: Confrontation or Consolidation?', Dushanbe/Brussels, 19 May 2004.

[67] Kambiz Arman and Nazar Nazarov, Eurasianet, 'Tajik President Seems to Resist Pressure for Political Liberalization', http://www.eurasianet.org/departments/civilsociety/articles/eav041905.shtml, accessed 23 August 2005.

[68] Nazar Nazarov, Eurasianet, 'Tajikistan Government Resorts to Repression and Intimidation', http://www.eurasianet.org/departments/insight/articles/eav042905a.shtml, accessed 13 October 2005.

[69] Human Rights Watch, 'US: Military Aid After 9/11 Threatens Human Rights', http://www.hrw.org/english/docs/2002/02/15/usint3750.htm, accessed 3 April 2005.

[70] Human Rights Watch, 'US: Military Aid After 9/11 Threatens Human Rights', http://www.hrw.org/english/docs/2002/02/15/usint3750.htm, accessed 3 April 2005.

[71] Human Rights Watch, 'US: Military Aid After 9/11 Threatens Human Rights', http://www.hrw.org/english/docs/2002/02/15/usint3750.htm, accessed 3 April 2005.

[72] Bureau of European and Eurasian Affairs, 'US Government Assistance to and Cooperative Activities with Eurasia', http://www.state.gov/p/eur/rls/rpt/37670.htm, accessed 24 June 2004.

[73] Bureau of European and Eurasian Affairs, 'US Government Assistance to and Cooperative Activities with Eurasia', http://www.state.gov/p/eur/rls/rpt/37670.htm, accessed 24 June 2004.

[74] Bureau of European and Eurasian Affairs, 'US Government Assistance to and Cooperative Activities with Eurasia', http://www.state.gov/p/eur/rls/rpt/37670.htm, accessed 13 December 2005.

[75] Bureau of European and Eurasian Affairs, 'US Government Assistance to and Cooperative Activities with Eurasia', http://www.state.gov/p/eur/rls/rpt/37670.htm, accessed 13 December 2005.

[76] Bureau of European and Eurasian Affairs, 'US Government Assistance to and Cooperative Activities with Eurasia', http://www.state.gov/p/eur/rls/rpt/37670.htm, accessed 13 December 2005.

[77] Bureau of European and Eurasian Affairs, 'US Government Assistance to and Cooperative Activities with Eurasia', http://www.state.gov/p/eur/rls/rpt/37670.htm, accessed 13 December 2005.

[78] Bureau of European and Eurasian Affairs, 'US Government Assistance to and Cooperative Activities with Eurasia', http://www.state.gov/p/eur/rls/rpt/37670.htm, accessed 13 December 2005.

[79] Bureau of European and Eurasian Affairs, 'US Government Assistance to and Cooperative Activities with Eurasia', http://www.state.gov/p/eur/rls/rpt/37670.htm, accessed 13 December 2005.

[80] Saodat Olimova, 'Political Islam and Conflict in Tajikistan', CA & CC Press, Sweden, 2005.

[81] Saodat Olimova, 'Political Islam and Conflict in Tajikistan', CA & CC Press, Sweden, 2005.

[82] Saodat Olimova, 'Political Islam and Conflict in Tajikistan', CA & CC Press, Sweden, 2005.

[83] Saodat Olimova, 'Political Islam and Conflict in Tajikistan', CA & CC Press, Sweden, 2005.

[84] Saodat Olimova, 'Political Islam and Conflict in Tajikistan', CA & CC Press, Sweden, 2005.

[85] Saodat Olimova, 'Political Islam and Conflict in Tajikistan', CA & CC Press, Sweden, 2005.

[86] Saodat Olimova, 'Political Islam and Conflict in Tajikistan', CA & CC Press, Sweden, 2005.

[87] Saodat Olimova, 'Political Islam and Conflict in Tajikistan', CA & CC Press, Sweden, 2005.

88 Olivier Roy, 'The Foreign Policy of the Central Asian Islamic Renaissance Party', Council on Foreign Relations, New York, 2000.

89 Olivier Roy, 'The Foreign Policy of the Central Asian Islamic Renaissance Party', Council on Foreign Relations, New York, 2000.

90 Olivier Roy, 'The Foreign Policy of the Central Asian Islamic Renaissance Party', Council on Foreign Relations, New York, 2000.

91 Olivier Roy, 'The Foreign Policy of the Central Asian Islamic Renaissance Party', Council on Foreign Relations, New York, 2000.

Chapter 6

Uzbekistan: A Cloud-Cuckoo-Land Place

Unlike in Kyrgyzstan, where the present work revealed a democratic revolution not all that democratic, or in Tajikistan, where a much-lauded secular-Islamic coalition was exposed as a fraud, there are no hidden surprises to offer on Uzbekistan. On the contrary, Uzbekistan has always been openly and sometimes stunningly brazen in its repression of Islam and denial of political pluralism. Indeed, it has long been a state where the executive administration has pursued imagined demons on every political corner. It is not surprising, therefore, to see that violent forms of Islamic fundamentalism have grown most quickly in Uzbekistan. This chapter reveals the depths of Uzbek repression, the continued use of US Wonka Vision with the local regime and the subsequent consequences created in the war against terror. Ultimately recent events showed just how ridiculous and compromised such policy left the US, for it is in Uzbekistan that one can see the futility of Wonka diplomacy.

Islam and the State – Making Lists and Taking Names

There have always been two major domestic policy directions in Uzbekistan, neither of which is very good for the development of democracy. The first was a form of authoritarian modernization slightly modeled on the 'Asian tiger' successes of South East Asia and China. It clearly was the preferred strategy in early-independence Uzbekistan and even today will dominate government press conferences. This policy was dominated by those who served as 'reformers' in Uzbekistan – English-speaking technocrats who dominated the Ministries of Foreign Affairs and Finance.[1] It was not a coincidence that these were the Uzbek officials that most visiting foreign delegations saw – they provided an acceptable face and exhibited a subtle understanding of the buzzwords and catchphrases the West found crucially important. These talking points included human rights, pluralism, federalism, and civil society. Of course this represented nothing of substance and reflected no real commitment on the part of Uzbekistan but that seemed to matter little to the Western diplomatic community.

The other direction, which was a more accurate reflection of political and economic reality, was best described as a type of neo-feudalism.[2] Here anti-modernism ruled the day and Western states were openly regarded as a threat to be handled, not allies to be embraced. This direction detested globalization and favored instead the 'Uzbek way', which was nothing more than code for personal politics over rule of law and feudal nepotism over meritocracy. It was the state divided into personal fiefdoms, with each incestuously tied to each other in mutually democratic-constraining ways. This degradation was evident in every sphere of societal activity but was especially prevalent in the state's policy toward Islam. From the very beginning Tashkent exhibited an arrogance of power stunning in scope.

Getting Hold of Islam

It was during Gorbachev's glasnost period of the late 1980s that saw a real flowering of Islam across Uzbekistan. Restrictions on religious practice were loosened and both the government and private organizations rushed to establish new mosques and madrassas.[3] Ironically Karimov, being a major figure in the Uzbek communist party at the end of the Soviet Union, undoubtedly had a large hand in bringing about this permissiveness toward Islam. He acted like the good Muslim when he assumed the presidency in 1991: he went on a pilgrimage to Mecca and symbolically placed his right hand on a copy of the Koran for his swearing-in ceremony as president. Events soon after independence, however, made Karimov rethink this strategy that involved Islam as a major component.

The early 1990s saw the rapid rise of Islam coincide with a collapse of state control over society and a quickly disintegrating industrial base. Having witnessed the eruption of a so-called secular-Islamic civil war just to the south in Tajikistan, Karimov and his team began to fear not only Islamic opposition in specific but political opposition in general.[4] Thus the first attempts at open Islamism were quickly suppressed in Uzbekistan. This was done through a broad and deep program where extreme laws were passed that made normal religious activities illegal. Religious instruction and education became intrusively monitored and restricted, forcing many groups underground and making them more sympathetic to radicalism. Policing was quite brutal and arbitrary, resulting in many people losing their freedom even though they had nothing to do with subversion or extremism.[5] In short, the Uzbek administration always regarded Islam as a rival to be defeated rather than a faith to be truly embraced.

It was this initial fear that made the government stomp early attempts to legalize an Islamic political party. Membership in the Islamic Renaissance Party was outlawed in 1992. The leader of the Uzbek chapter of the IRP, Imam Abdulla Utaev, disappeared later that year and was never heard from again.[6] In less than six months Karimov had gone from basing the symbolic legality of his presidency on the Prophet Muhammad's word to showing a slightly paranoid fear of the general power of Islamic faith.

These efforts did not stop small informal Muslim groups from sprouting up all over Uzbekistan like mushrooms, however. This was especially so throughout the Ferghana Valley, which revitalized its historical image of being a center of Islamic faith.[7] In Namangan, one of Ferghana's major cities, one emergent group was called Adolat. Basically a young men's Islamic militia which sought to fill the power void after the fall of the Soviet Union, it enforced Sharia law over the local population through intimidation.

Adolat was not a revolutionary group aimed at taking over the country and creating an Islamic state. More than anything, it functioned like a rather severe citizen's watchdog group that would walk the streets and mete out swift judgment and punishment on hooliganism and petty crime. While the group clearly leaned too heavily on draconian punishments (one favorite was to lead thieves and prostitutes through the city, sitting backwards on donkeys, while citizens were encouraged to come out and spit at them), many citizens long after would wax nostalgic about how it was possible during 'Adolat's reign' to leave car doors unlocked and windows open at night. Crime virtually disappeared.[8]

What became of Adolat? Their efforts at official recognition from Karimov gave them more than they bargained for. Karimov at first so supported Adolat that he visited Namangan in early 1992. He was greeted, however, by crowds of Islamist supporters demanding Adolat's legalization. Karimov did not anticipate such passion. He thought he was entering the Ferghana Valley as an adored leader. Instead he left Namangan shaken and disturbed at no longer being the master across all of his own territory.[9] Before the year was out Adolat and all small informal groups like it were outlawed. Many of the original Adolat members were arrested and given long prison sentences.[10] Those who were successful in evading arrest fled to Tajikistan and then moved on to Afghanistan to form the core of the bloody and revolutionary Islamic Movement of Uzbekistan (IMU). Thus from the outset, a clear and dangerous path was drawn between government oppression and Islamic fundamentalism. In fact this empirical example vividly shows the consequences to such underhanded dealing with what could be potentially moderate Islamic groups: if they cannot be legally moderate then they will inevitably become illegally radical.

The widespread stigmatization of what the administration deemed 'Wahhabists' was a key element of Karimov policy. In reality, they were usually not like the Wahhabis found in Saudi Arabia but were traditional Islamists practicing a formal and conservative form of Hanafi Sunnism, which was always common across Uzbekistan. Pushing the stigma forward was a 1993 law on religion that eerily copied Soviet censorship. It created two governing bodies, the Muslim Spiritual Board and the Cabinet of Ministers' Committee on Religious Affairs (CRA), which were charged with first defining acceptable Muslim practices and then using those definitions to weed out violators.[11]

These violators were subsequently branded 'Wahhabists' and marked for repression. To this day the government uses the word Wahhabist as an epithet signifying 'anti-Uzbek'. In reality, this 'anti-Uzbek' behavior usually involved imams refusing to pay homage to President Karimov during prayer hours or refusing to become government informants and spying on their own congregations. To Karimov these were considered acts against the 'constitutional' order.[12]

The CRA controlled the Muslim Spiritual Board, which in turn was responsible for registering mosques and madrassas, appointing and dismissing individual imams, dictating the contents of sermons, and issuing religious edicts.[13] The CRA was not an actual decision-making body. In fact it was a rubber stamp for President Karimov. The main weapon in the arsenal of the CRA was registration. This was a labyrinthine process pregnant with corruption. Officially, registration was done through the Ministry of Justice with the approval of the CRA. In reality protests against the registration of a particular mosque or madrassa could literally come from any level – from local municipalities to the President himself – making it extremely difficult to pass all the registration checks.[14] More importantly, rejection of a registration application did not require a formal explanation.

The religious hierarchy was basically a mimic of Soviet-era structures. The muftiate was headed by an individual with little real societal authority but who remained extremely loyal to the state.[15] The Muslim Spiritual Board was thus subverted under the CRA and used to control Islam, mainly in one of two ways. The first was through the appointment of imams, which was done through a bulky and awkward 'negotiation' between local authorities, security services and the CRA.[16] In Uzbekistan it was clear that successful appointment came only with the ability to convince the governing bodies that an imam's political views were properly subjugated and in line with the executive administration. This ended up tainting the entire religious hierarchy with illegitimacy and political compromise, best described by a local Muslim when asked about the process:

> I studied in the madrassa myself and know how imams are chosen here. The main criterion is loyalty to the government and readiness to pay bribes and give gifts to the Muftiate and state officials. Knowledge and morality does not interest them. Corruption and nepotism has moved from state structures into the sphere of religion.[17]

This work revealed the same problem in Kyrgyzstan and Tajikistan but it is in its fullest severity in Uzbekistan: government co-optation is so complete that it drives people away from moderate Islam, which has become synonymous with corruption and immorality. Moderate Islam, Uzbek-style, is the absence of sincere faith not proof of harmony between secular and religious authorities.

The second form of control available to the muftiate was the official distribution of information, usually done through fatwas and sermons.[18] Imams did not deliver sermons on subjects of their choosing. That was a quick and easy cause for removal. Instead they were provided with the basic themes of each weekly sermon by the muftiate, which first passed the approval of the CRA, which was itself a mouthpiece for Karimov. The president in this way sort of writes the very sermons believers hear all over Uzbekistan. As a consequence, 'official' Islam in

Uzbekistan never addressed the most pressing and important societal needs.[19] Instead the faithful got a watered-down version of Islam combined with a praising of Karimov policy. In this surreal world of 'religious freedom' only one type of group truly benefited: radical Islamists, as they were the only groups in Uzbek society that truly provided something close to real Islamic philosophy and unfettered criticism.

The government extended its reach beyond the mosque by also controlling Islamic study, instruction and schooling. The sometimes extreme suppression of orthodox Islamic teaching only succeeded in pushing ever more talented scholars underground. Over time this control only deepened, most notably with an amendment to the 1999 law on religion. The amendment outlawed the private teaching of religion and also banned teaching in mosques unless it came under the specific permission of the muftiate. This was enforced by the Ministry of Internal Affairs by forbidding people from staying in mosques after the conclusion of formal prayers.[20] Thus in every way possible the Karimov administration managed to subordinate official Islam to its political whim. This came with a price, however: there was deepening societal discontent and large segments of the population became radicalized because moderate Islam was seen as a non-choice. It was a vicious cycle: fear of radicalism led to repression which converted more to radicalism which justified greater repression. Karimov was not shy about ratcheting up the intensity. After all, this was the president who before an open microphone declared that Islamic extremists were 'such people who should be shot in the head. If necessary, I will shoot them myself'.[21]

Making Imams the Enemy

The disappearance of Sheikh Abduvali Kori Mirzoev was the first in an unfortunate series of 'unexplained events' that signaled a harshening of government oppression in Uzbekistan. Sheikh Mirzoev was the chief imam of the Jo'mi (Friday) mosque in Andijan. He was clearly one of the most outspoken of the 'authoritative imams', who advocated the establishment of an Islamic state based on sharia and consistently resisted government efforts to control his religious services and sermons.[22] Not coincidentally, Mirzoev was one of the most revered spiritual leaders in all of Uzbekistan. Human Rights Watch best described the strange events leading up to his disappearance.

> On 29 August 1995, Mirzoev and his assistant, Ramazanbek Matkarimov, were scheduled to fly from Tashkent to Moscow to attend an international conference. There were reports that one eyewitness at the airport saw security agents detain Mirzoev as he boarded the plane. There were no confirmed reports of sightings of either Mirzoev or his assistant afterward. The Uzbek government, for its part, alleged that Imam Mirzoev had fabricated his own disappearance in order to justify the claim that the government was violating independent Muslims' freedom of religion. The Jo'mi mosque, which the government had labeled a Wahhabi mosque and a source of reactionary religious ideas, was closed after

Mirzoev's disappearance in 1995…[The fate of Mirzoev and Matkarimov was never ascertained.] Mirzoev's former assistant, Nematjon Parpiev, followed in the Sheikh's footsteps to become an imam in Andijan. Parpiev disappeared in September 1997.[23]

When the government wasn't busy creating its own conspiracy theories to evade culpability in the obvious murder of imams, it was using the Muslim Spiritual Board to threaten, harass and dismiss imams at will. Among these was Imam Obidkhon Kori Nazarov of the Tokhtaboi mosque in Tashkent.[24] Nazarov was actually rumored to be in line for the position of mufti, making him occupy the highest post in the official Uzbek Islamic clergy. By 1996 Nazarov was balking at the Spiritual Board's incessant interference in the content of his sermons and refused to serve as an informant for the National Security Service, the Uzbek successor to the KGB. As a consequence, the Spiritual Board dismissed Nazarov from his post for 'disobedience to the Board'.[25]

Bombs Bursting in Air: Terrorist Actions and Widening Arrests

All of these actions did not take place in a vacuum nor did Uzbek authorities succeed in making society cower completely. Terrorist retaliations and initiatives did take place throughout Uzbekistan and especially in the Ferghana Valley where operatives found it extremely easy to hit targets and then melt into the background. The 'anti-Wahhabi' initiatives increased in intensity after a tragic incident in 1997 when several police officers were murdered and two others, a local government official and another prominent community member, were beheaded in Namangan.

The government wasted little time in arresting *several hundred* people. Within four months the Ferghana strongholds of Namangan and Andijan provinces had more than one thousand people arrested. The authorities simply arrested all those even remotely affiliated with imams Nazarov and Mirzoev. The murders and beheadings gave justification to a massive increase in surveillance of Muslim organizations. Beginning in 1998 intelligence agencies, local administrators, and mahalla committees were pressured by the government to 'monitor' suspicious religious activity.[26] The problem was that the Karimov government basically considered anyone religious as a source of anti-government fervor. By 1999 Karimov said openly that mahalla committees should be especially alert so as to quickly determine what could be considered improper behavior.[27] In short, Karimov was creating a system that by design encouraged and rewarded citizens to inform on one another. This intrusive system resulted in the production of 'lists' that ran on for thousands of names. All of these people would ultimately be declared 'enemies of the state'. When Tashkent was hit by a series of bombs in 1999, many on the lists after the 1998 murders were summarily arrested even though the vast majority had no possible connection to the bombings.

On 16 February 1999 five bombs exploded near government buildings in Tashkent, claiming more than a dozen lives and wounding countless others. President Karimov obviously accused Islamic extremists and began another massive round of arbitrary arrests and intimidation.[28] Security services arrested literally thousands of Muslims. President Karimov, who was personally embarrassed at the success of a terrorist operation in the heart of the capital, considered the bombings an attempt on his life and vowed to hold accountable 'the fathers who have brought up enemies of the state, together with their children'.[29] As if executive paranoia, societal informants, imam spying and arbitrary arrests were not bad enough, the Tashkent bombings basically ended up making 'hereditary punishment' legal in Uzbekistan. Not immediately noticed by Western analysts in the middle of all of these bombings, beheadings and bad arrests was a watershed legal moment in Uzbek politics. In many ways, Uzbek society still suffers from the consequences created with the passing of the 1998 Law on Freedom of Conscience and Religious Organizations.

The Law on Freedom of Conscience and Religious Organizations

It was this 1998 law project, passed in conjunction with complementary amendments to the criminal and administrative legal codes, which signaled to Uzbek society that the government intended to take a systematic approach to annihilating true religious freedom. Up to that point there had always been fall-back excuses that blamed other sundry whipping posts: religious oppression was the fault of the corrupt bureaucracy or police were so poorly paid that they needed to violate the law in order to earn a proper living. The 1998 law revealed the true nature of Karimov, who had no desire to see a sincere political Islam operate within the territory of Uzbekistan. When the various articles and amendments that accompanied the law are examined, the totality of oppression is stunning.[30]

- Article 5 criminalized 'actions aimed at converting believers of one religion into another (proselytism)'. Accompanying this article was criminal code article 216–2, part 2, which criminalized any conversion of believers of a faith.

- Article 9 criminalized private religious teaching and instruction on religious principles. Accompanying this article was criminal code article 229–2, which criminalized 'teaching religious dogmas without special religious education and without permission of the Central Administrative Body of a given Religious Organization, as well as teaching religious dogmas in private'.

- Articles 8 and 11 criminalized unregistered religious association by specifying that only registered religious organizations have the right to function as legal entities. Accompanying these articles were criminal code article 216–2, part 1, which pertained to the failure to register a religious organization, articles 216 and 216–1, which referred to the organization of, participation in and persuading of others to join a prohibited religious organization.

- Article 19 criminalized the production, storage and distribution of materials that 'contained ideas of religious extremism, separatism and fundamentalism'. These materials included printed documents, video and audio cassettes, films, and photographs. Accompanying this article was criminal code article 244–1, which criminalized the possession and distribution of literature containing ideas of 'religious extremism, separatism, and fundamentalism'.

- Criminal code article 244–2 made religious, extremist, separatist and fundamentalist association illegal. In fact, any actions that could be considered 'setting up, leading and participating in a religious extremist group' was strictly forbidden.

- Article 14 prohibited the open wearing of religious dress, specifically outlawing the wearing of religious clothing by non-clerics. Administrative code article 184–1 accompanied this by specifically elaborating the penalties incurred for violations of Article 14.

Attempts to actually observe the law were usually frustrated as the minutiae necessary to be in accordance was laughable. Understand that failure to register an organization under Uzbek law did not mean the organization simply lost certain commercial rights. It meant that the group itself was illegal and therefore *all of the members* of the group were themselves subsequently considered criminals.

Out of all the accompanying criminal code amendments adopted to facilitate the repression against Islam, articles 159 and 156 were clearly the most instrumental. Article 159 was entitled 'Encroachment on the Constitutional Order of the Republic of Uzbekistan' and gave prosecutors a weapon applicable to all circumstances. Article 156 referred to the government's right to prevent the 'incitement of national (ethnic), racial, or religious enmity' within Uzbekistan. In both cases the articles proved especially effective in being able to create Muslim opposition out of thin air.

> Article 159 – Public appeals to unconstitutionally change the existing governmental system, to seize power to remove from office legally elected or appointed representatives, or to unconstitutionally disrupt the territorial unity of the Republic of Uzbekistan, as well as distribution of material with such content are punishable with a fine of up to fifty times the minimum wage or imprisonment up to three years.[31]

Most important to note was the fact that Article 159 was irrelevant to *actual* radical Islamist groups. There was little the Karimov government could do legally to deter radicals from fighting. Passing a law was not going to make revolutionaries desist. Therefore it seems obvious that Article 159 was not aimed at radicals but at their families and friends. To a rebel bent on overthrowing the regime or a radical supportive of civil unrest a heavy fine and/or three years in prison were not likely to impress, but to a regular citizen the fine alone guaranteed de facto indentured servitude to the state. The articles ostensibly gave security agents a permanent legal reason to arrest and detain any imam or religious figure that crossed the administration. This was why most major imams and Islamic leaders in Uzbekistan intermittently spent time in prison.[32]

When viewing the articles in sum what appears is a system of jurisprudence that annihilates not the *practice* of religion per se but its actual *existence* in society. What exactly did the above articles outlaw? They outlawed conversion, teaching, proselytizing, association, information (including its creation, distribution and even possession), and expression (in this case, physical dress). If these six aspects are removed from a person of faith and criminalized so their presence in a person's life can lead to imprisonment then what exactly of religion remains?

Allah, Islam, and Muhammad all ceased to legally exist with the passing of the 1998 law. Putting it in any other way would simply be avoiding the truth of the matter and the true essence of the law. Karimov was not trying to destroy fundamentalism. He was waging a war against Islam in all of its possible forms. Such initiatives destroyed any chance that a truly moderate Islam would develop in Uzbek politics. The only refuge available to an Uzbek who wished to participate in his country's politics and simultaneously stay loyal to his broader religious faith was radicalism. Karimov therefore should be seen not as reacting to fundamentalism: he *caused* it. The Tashkent bombings happened *after* the passing of the law because to radical Islamists he had already openly declared war.

Turning the Screws: Waging War Against Islam

By summer 1999 a series of trials began not so much to find and punish the perpetrators of the Tashkent bombings but to send a signal to the rest of society that may have felt sympathy for the acts. Defendants were only referred to as 'Wahhabis' and were continually implicated in a wider conspiracy involving thousands of people intent on taking down the government. Prosecutors provided no material or forensic evidence during the trial and simply filibustered on their various conspiracy theories.[33] Behind the courtroom drama was the routine torture of defendants: holding them in communicado, depriving them of legal counsel and intimidating both witnesses and relatives into testifying on the government's behalf at the trials. Usually the most damning 'evidence' came from members of the Committee on Religious Affairs (CRA), who nevertheless did not appear in person at trial nor were their reports ever made available to defense attorneys for analysis.[34] In fact, at every trial independent analysis of the 'incriminating evidence' was almost never allowed. Content was never discussed openly and usually only the title of an offending leaflet or booklet was mentioned.[35] In other words, Karimov blatantly and brazenly created a 'show trial' environment rivaling anything during the Stalin era.

Many of the people who had previously made government lists for potential extremist sympathies found themselves once again rounded up and arrested. As a consequence, Uzbek jails quickly filled to capacity with so-called 'extremists' and sympathizers. It was this reality which cynically caused Karimov in 1999 and again in 2002 to offer a magnanimous 'amnesty' to all those who wished to turn themselves in and receive a pardon for cooperation with the government.[36] The official procedure to receive the pardon required citizens to ask the state for forgiveness for engaging in unsanctioned and therefore illegal religious activity:

> Those convicted and sentenced for the first time to prison terms of up to ten years inclusive of who were given penalties other than imprisonment for involvement in the activities of extremist organizations and committing, as a member of these, a crime against the constitutional order of the Republic of Uzbekistan or other activities against public security.

The real point behind the amnesty was to root out more extremist sympathizers while actually freeing up jail space which had become occupied by regular citizens arbitrarily and falsely imprisoned. Parents were encouraged to expose their own children in expectation of helping them receive more lenient treatment. In addition, it was clear many who felt their families and friends would fall under duress and intimidation went in voluntarily in order to secure not their own safety but the safety of their loved ones.

Hundreds of those even slightly affiliated or connected to 'extremist' organizations were tried and imprisoned, *even though they emerged in order to receive the amnesty*. The amnesties were nothing but a revolving door that released some to accommodate others.[37] Once the 'confessed extremists' were convicted they were taken on a surreal tour of Uzbek regions and paraded in front of neighborhoods at public events in order to drum up animosity against their families.[38] It wasn't only the confessors who were obligated to attend these denunciations but even the public was forced by local officials. Failure to attend the hate rallies inevitably made those not participating themselves objects of scrutiny. The deception and deviousness was amusing if not for the suffering incurred and the baffling totality of Karimov's Big Brother world.

Police in the Karimov system rampaged through society with impunity. Not only was it a rare exception for police to properly identify themselves, they often actively sought to cloak and disguise their identity from citizens. The laws on search and seizure were never respected and often turned inside out as the planting of evidence, sometimes in full and open view, became notoriously commonplace.[39] Under the legal system devised by Karimov a simple leaflet or insignificant booklet was just cause to arrest entire families.

For the most part Karimov's main demon was the Islamic Movement of Uzbekistan (IMU), led militarily by Juma Namangani and politically by Tokhir Yuldash. While it was true that the IMU was an armed revolutionary group committed to the violent overthrow of the Karimov regime, it was also clear that the administration used the threat of the IMU to broaden and deepen the scope and reach of its own repressive measures. Both the judiciary branch and procuracy had legal obligations to prevent detainees from cruel and unusual suffering. In real terms, however, torture was rarely investigated, often only on account of the fact that someone was actually tortured to death and a body needed to be explained.[40] When it came to torture, Jaslyk prison stood out. Located in the desert of Karakalpakstan Autonomous Republic in the far northeast of the country, the prison became legendary for high rates of torture, prisoner death and systemic mistreatment by guards.[41] While nearly impossible to confirm exact numbers, it was clear that Jaslyk during its heyday held at least several hundred prisoners. Attorneys were basically not permitted to reach Jaslyk and family visits were only allowed on extremely rare occasions. In fact, Jaslyk served the Uzbek penal system as a type of 'black hole'. Authorities routinely transferred prisoners from other prisons to Jaslyk without notifying family members of the switch. In a state with a well-deserved reputation for penal brutality, Jaslyk was the crown jewel, dedicated entirely to detaining religious and political prisoners.[42]

Even the Uzbek institution meant to serve citizens' rights in terms of redress and reparation for unlawful prosecution, the Ombudsman office, was a formal illusion and executed its duty in total hypocrisy. Uzbekistan's first Ombudsman, Sayora Rashidova, actually filed a public report in 2000 on Jaslyk prison expressing her overall *satisfaction* with the facility.

> Conditions created for prisoners [and the fact that] the corps of guards were observing the law in their duties made us happy…I can say that the supply of food and clothes for them is in good condition. What else can a person need who was found guilty by law and who had admitted to [crimes?] Is the fact that such conditions have been created for prisoners not enough to state that human rights in Uzbekistan are being observed at every step?[43]

This was filed by the office actually created by Karimov as an *appeasement* to international outcries against Uzbek repression. Indeed, it was only a few years later that the British Ambassador to Uzbekistan forensically confirmed the deaths of two inmates at Jaslyk prison by being brutally beaten and then boiled alive. The only reason their deaths came to light was the fact that the mother of one of the tortured, Fatima Mukhadirova, did not remain quiet and publicly sought justice for her dead son. In fitting symbolism, Mukhadirova herself was charged and sentenced to prison for 'religious extremism and plotting against the state'.[44] It was only with the impending visit of US Secretary of Defense Donald Rumsfeld to Uzbekistan in 2004 that Mukhadirova was quietly released.[45]

This systematic repression received little official condemnation from foreign governments. Some human rights groups were striving to draw more attention to the Karimov regime but most foreign governments were indifferent. Except for the US, which just after 9/11 literally stood in unison with Karimov and made its own conspiracy connection, tying the IMU to the Taliban and to hidden Al-Qaeda bases in northern Afghanistan. Thus part of US operations in Afghanistan in late 2001 were not against encampments of Al-Qaeda fighters or Taliban regiments but actually against IMU fighters. It was such an operation that killed Juma Namangani. This was but one example of the direct assistance the US gave to Karimov. The so-called tightrope policy of condemning human rights abuses on the one hand while urging reforms and staying 'engaged' on the other did nothing but allow Uzbeks to be tortured and radical Islamists to intensify and expand their hatred.[46]

Return of the Wonka: US Diplomacy in Uzbekistan

Perhaps the 'tightrope' moniker is not ideal in describing US policy towards Uzbekistan. 'Pendulum' is probably better as from the beginning the United States has always softly condemned human rights abuses or the lack of progress in democratic development while also being quick to point out supposed advancements 'on account of US engagement'. This pendulum swung much less toward criticism after 9/11 because it was clear to both the Bush administration and Karimov that Uzbekistan was the best Central Asian state to enlist as an ally in the war against terror. Only Uzbekistan had a president with enough executive power and hatred of 'radical Islamists' to be eager to engage in the 'Global War against Terror'. The potential dire consequences to such friendship were clear but ignored: since Karimov did not care that much about international opinion before 9/11, when his country was basically an international diplomatic backwater, it was not difficult to guess his attitude after 9/11 when his state became a major lynch pin in the larger US war.

Less than three weeks after 9/11 Defense Secretary Rumsfeld was in Uzbekistan meeting with Karimov. At the press conference Rumsfeld was in full spin mode:

> We concentrated on issues pertaining to counter-terrorism and elimination of the mechanism that we call terrorism…We're not fighting against any one single country, not against the Taliban for that matter, but we're fighting against camps and bases and infrastructure worldwide…
>
> The first point is Uzbekistan grants its airspace to American aircraft and helicopters. The second point is Uzbekistan is ready to upgrade and step up cooperation between special services for the exchange of intelligence information. Uzbekistan gives its permission and gives use of one of its airfields and its facilities for United States' aircraft and helicopters as well as for personnel employed in search and rescue operations…
>
> It should be said that in the course of these three years, Uzbekistan has been witnessing the inhumane face of terror; therefore we cannot afford standing aside and we are taking part in this anti-terror operation that the international community called for…We benefit not only by the cooperation, but also by the insights and the perspectives that the president offered…
>
> The interest of the United States in Uzbekistan, it should well be understood, precedes the events of September 11…And the interest of the United States is a long-standing relationship with this country and not something that is focused on the immediate problem alone.[47]

Putting aside the gross exaggeration of US interest in Uzbekistan being more than just about the immediate problem of terrorism, Rumsfeld was basically admitting that the Uzbek partnership was just because of its convenient geographical location and its ability to gain and share intelligence on terrorists. This chapter has detailed the brutal ways that Uzbekistan gained intelligence. Therefore while Rumsfeld's statement came off fairly banal to the West it sounded to the local community like he was signaling US approval for such brutality and the willingness of the United States to profit from it.

It was not only Rumsfeld who made embarrassing trips to Uzbekistan. Treasury Secretary Paul O'Neill also did in July 2002. His statements were so egregiously off-fact that his press conference with President Karimov was and still is used for propaganda purposes by radical Islamists:[48]

> O'NEILL: It's a great pleasure to have had this opportunity to meet with the President and with the Deputy Prime Minister and with some of the other ministers to explore issues and particularly to thank the President and the people for their support for the war on terrorism…
>
> I expressed to the President our admiration for the leadership that he has provided during the economic transition, giving a very high priority to education and the important human needs of the people of Uzbekistan. It's a great pleasure to have an opportunity to spend time with someone of both a very keen intellect and a deep passion about the involvement of the life of the people of this country…
>
> QUESTION FROM REUTERS: How satisfied are you that the momentum toward the type of economic reforms that America wants for Uzbekistan is being maintained?
>
> O'NIELL: I came away assured that the President and his people are moving with deliberate speed with no doubt about policy goals and objectives, and I don't mean by that they are moving slowly. I do mean that they are moving as quickly as they know how to move…
>
> QUESTION FROM UZBEK-TV: Mr. O'Neill, what are the economic and financial aspects of the Uzbek-American strategic partnership? Are there concrete projects and in what fields?
>
> O'NEILL: Well, I think we have a very strong partnership and it's certainly been strengthened by our work together since 9/11.
>
> QUESTION FROM THE BBC: President Karimov, with the apparent struggle between reformers and more conservative elements within your government, will it be possible for Uzbekistan to carry out the reforms?

> KARIMOV: I would like to say that no society, no country in the world has gone through these processes without pain…Today during our talks I said to Mr. O'Neill that for Uzbekistan the hardest question is the reformation of our society, the democratization of our society, building a civil society, and raising political activity and awareness of our population. It is very important that we overcome the vices inherited from the totalitarian system, from the administrative-command system that we lived with for more than 75 years…
>
> I would like to say unambiguously that whatever is being done in Uzbekistan today and the achievements we have had for the past ten years and especially during the last three prove we are moving forward along the road of democracy.
>
> O'NEILL: I quite agree with the President…It would be well for you to go and make inquiries of people across the region who have realized the difference from living in a welfare state or one that's controlled by totalitarianism to the freedom that they feel when they have their own land, and they own their own land and they have the ability to create a good life for themselves and their families.
>
> Maybe I haven't been looking in the right places, but I haven't found anyone who has had the opportunity to become a land-owner instead of a slave to a government who prefers to be a slave to a government.

It was not surprising to hear Karimov's bluster or his rationalizations but it was shocking to find how off-base O'Neill's understanding of Uzbek political reality was. Indeed, the rhetorical dismissiveness with which he ended the press conference was simply embarrassing: in Uzbekistan people are prevented from creating a good life for themselves and their families; in Uzbekistan people do not choose between Soviet totalitarianism and liberal democracy, they choose between a corrupt kleptocracy and an oppressive political dictatorship. They can only choose between these options *because* of the man that was standing next to O'Neill. His ignorance of that reality did not earn sympathy or understanding from radical Islamists. O'Neill's ignorance was so far beyond believability that most took it to be done purposefully. It seemed obvious to radicals and moderates alike that the fate of the majority of Uzbeks were a horrendously low priority to the United States.

More than two years later, on another visit to Uzbekistan, Rumsfeld showed a deft touch in avoiding difficult questions but came off looking so insincere that it did nothing except prove that there had been no real shift in US policy since 9/11. Closer relations with Tashkent did not produce closer scrutiny. The Uzbek people were left to fend for themselves.[49]

> RUMSFELD: Uzbekistan is a key member of the coalition's global war on terror. And I brought the President good wishes of President Bush and our appreciation for their stalwart support…Today we discussed the excellent military-to-military relationship between Uzbekistan and the United States. Our

relationship is strong and has been growing stronger…The relationship between our two countries is important to us, and we certainly value the friendship of Uzbekistan and the people of Uzbekistan and are grateful for the stalwart, steadfast support in our efforts against terrorism…

REUTERS: You spoke of this strategic framework, of the relationship between two countries. Uzbekistan said yesterday they're going to free a 62-year old woman from jail, who human rights activists say was jailed on trumped up charges because she revealed her son had been tortured to death in prison. Do you welcome this, sir, and to what extent will improvements in human rights in this country deal with continued US military aid to Uzbekistan? (This is referring to the earlier covered death by boiling, see page 129).

RUMSFELD: Well, obviously our relationship with this country and other countries is multi-faceted. I mentioned the military-to-military relationship because I'm involved with the Department of Defense, but it's also a political and economic relationship. I'm not intimately knowledgeable about the statement you just made, but my Embassy has expressed their awareness of that and I forget what the phrase was but…the Ambassador pointed out that they were pleased that the decision was made.

BBC RADIO: To what extent the assistance provided by the US government to Uzbekistan is linked to the human rights issue and the release of Mrs. Makhadirova, which has been mentioned today, was the mother of the person who was in detention in Jaslyk camp, who actually died, and the scientists from Glasgow have confirmed that the person died as a result of serious torture. To what extent have you heard about this?

RUMSFELD: I'm not knowledgeable about every aspect of this. The Ambassador has responded that the United States is pleased with the release that's been made. And the answer is that the relationships between sovereign nations tend not to be made on a single pillar.

It was disingenuous, at best, for Rumsfeld to deny sufficient knowledge of the boiling torture incident as it had become something of a cause celebre in the human rights community for the past year and even resulted in the British Ambassador to Uzbekistan ostensibly being fired because of his own vehement protests against the barbarity. Worse still was the ultimate indifference Rumsfeld showed to the plight of the Uzbek people by claiming sovereign nations build relations on more than just one pillar. Indeed, this is undoubtedly true but the boiling torture incident cut across numerous pillars: it signaled the absence of political openness, the annihilation of due process, the corruption of the judiciary and the procuracy, the breakdown of the criminal justice system and the degradation of the holy principle of all sovereign governments – to first protect their own people. ALL of those pillars were weakened if not destroyed by the boiling torture. Rumsfeld's take that it should be considered an isolated incident remedied by the release from prison of the mother of the deceased was preposterous.

Only one month after the above press conference a series of suicide bombings took place across Tashkent and Bukhara, claiming 45 lives. Islamic Jihad, a group previously unknown in Uzbekistan, claimed responsibility, pinning the attacks on Uzbekistan's support of the US war against terror.[50] For the most part many analysts were surprised by this development. Unlike previous terrorist attacks in Uzbekistan this one carried all the hallmarks of truly belonging to the 'global jihadi network':[51]

- Lengthy lead times to indoctrinate attackers
- Long preparation for operations
- A large number of perpetrators
- The use of women suicide bombers, as in Chechnya, Gaza and the West Bank
- The lack of a stated political platform
- The indiscriminate targeting of civilians, including women and children.

Despite acknowledging the blatant repression of Karimov's domestic policies there was still a general consensus in policy circles that Washington had to exercise care in being more openly critical, 'lest Uzbekistan be destabilized'.[52] The excuse of 'no clear democratic alternative' reared its ugly head as Karimov was still regarded as the only man capable for real negotiations. Some analysts were so bold as to criticize the 'liberal-left human rights community' for not properly understanding the Islamist terrorist threat that made continued engagement with Uzbekistan a better alternative than forceful criticism.[53]

This policy-wonk consensus missed the most important realities of Uzbek politics. By 2004 what was once a purely domestic terrorist threat in Uzbekistan had already begun to be transformed into part of the global terrorist network. This transformation did not take place in spite of US engagement it took place *because* of it. The ideal of engagement, of productive negotiations garnering reform concessions, was noble but blatantly insincere: this type of engagement never took place in Uzbekistan. Instead, when analyzing how our assistance and engagement was formally judged, a system of unfathomable ignorance and moronic hypocrisy emerged.

As was documented in the Tajikistan chapter, all US government assistance to foreign states has to pass section 489(a) of the Foreign Assistance Act of 1961. When looking over the official report released from the Bureau of European and Eurasian Affairs on Uzbekistan in 2004, Wonka Vision diplomacy was so ingrained that parts of the report actually *contradicted* themselves. The only mystery left to readers is to determine whether this contradiction was done purposefully or simply indicated a severe level of incompetence across the entire American diplomatic bureaucracy.

> Section 489(a)(1): make significant progress toward, and is committed to the comprehensive implementation of, a democratic system based on principles of the rule of law, individual freedoms, and representative government determined by free and fair elections.
>
> - While the overall human rights and democratic situation remains very poor in Uzbekistan, there have been small steps in the area of democratization and rule of law in the past year.
>
> - The Constitution of Uzbekistan provides for the separation of powers between the executive, legislative and judicial branches. However, President Karimov and the centralized executive branch that serves his authoritarian regime dominated political life.
>
> - The government held an advisory referendum on extension of the term of the president from five to seven years. The referendum passed by a large margin and was judged to be neither free nor fair by most observers. Parliament then passed a constitutional amendment extending the term. President Karimov announced that the next presidential elections would take place in December 2007, resulting in an extension of his current term to just short of eight years.[54]

Not to state the obvious, but exactly how did any of this signify even the tiniest of small steps in the area of democratization and the rule of law?

> Section 489(s)(3): Respect internationally recognized human rights, including the rights of minorities and the rights to freedom of religion and emigration.
>
> - Although the overall human rights situation remains very poor, the government has taken some small but significant steps to improve the situation in the past year.
>
> - Uzbekistan authorities continued to commit numerous serious abuses, such as torture of detainees.
>
> - Uzbekistan's citizens are still prevented from exercising their right to change their government peacefully.
>
> - Uzbekistan's constitution provides for an independent judiciary. However, the judicial branch takes its direction from the executive branch and has little independence in practice.[55]

Again, if these small but significant steps existed they were certainly dwarfed by the even larger steps that refuted the principles of democracy. The United States recommended to the Uzbek government six specific steps to be taken to demonstrate a commitment to human rights. These steps were required so as to receive assistance under the Cooperative Threat Reduction Act.[56] Uzbekistan did not do a single one:

- President Karimov did not publicly condemn torture.
- There was no independent forensic investigation or accountability for the two torture deaths in May or another death in December.
- Although the Uzbek government took the 'admirable step' of proposing a draft action plan to combat torture in response to the visit of the UN Special Rapporteur on Torture, who concluded that torture in Uzbekistan was 'systematic', the proposals were not implemented.
- The Uzbek government did not register any opposition parties.
- Uzbekistan did not introduce election legislation to be compliant with OSCE standards.
- Uzbekistan did not make a list of all prisoners released in the most recent amnesty.

As to the US reaction in the face of such open defiance? The Bureau issued a soft entreaty, mentioning that 'an intensive and continuing dialogue concerning the need to address human rights abuses has achieved some important, but limited, results'.[57] Most importantly, Uzbekistan still received funds under the Cooperative Threat Reduction Act. Its failure to enact *any* of the recommended steps made it legally ineligible until President Bush signed a special exemption.

In fact, Wonka Vision diplomacy seemed to be worse in Uzbekistan, the Ferghana state with easily the most repressive record, than in either Tajikistan or Kyrgyzstan, where it was admittedly bad. Continuously the Bureau's reports referred to engagement and dialogue. Embarrassingly, the US government more often than not simply looked overjoyed whenever able to declare 'favorable responses, intensive dialogue and modest results'.[58] This was nowhere more apparent than in Section 489(b)(1):

> Has the President determined that the government of Uzbekistan has engaged in a consistent pattern of gross violation of internationally recognized human rights or of international law?

- No. The President has not made such a determination at this time. However, the US government remains concerned by the human rights situation in Uzbekistan. Lack of freedoms of association…and torture raise serious concerns regarding the credibility of the system of justice in Uzbekistan…Despite these problems, the government of Uzbekistan has taken small but significant steps to improve its human rights records.[59]

A careful examination of Uzbekistan's political evolution reveals no small steps. It reveals only a big brother society that would shock even George Orwell and an international partner in the United States that seems blissfully oblivious.

Uzbek Politics: Power Incorporated

Uzbek presidential and parliamentary politics have served one master since 1991: Islam Karimov. It is difficult to find a single decision, initiative or referendum since that time that did not in fact increase the breadth and depth of Karimov's power. He first gained 'legitimacy' in presidential elections at the end of 1991 when his victory was declared at 86% of the total vote cast. It was undoubtedly rigged. In fact, it was only in this first election that there were actual opposition figures that legitimately opposed Karimov. He made sure that such a mistake did not happen again.

Though elections were scheduled for 1997, members of the newly elected parliament – the Oliy Majlis – adopted a resolution in February 1995 calling for a referendum to prolong Karimov's term until 2000. In March 1996 supposedly 99.6% of the population voted for this referendum, giving Karimov a first term that ostensibly extended eight years and also garnered him the right to stand for president yet again.[60] Following the referendum victory Karimov officially withdrew from the NDPU as he pushed for legislation that demanded a president *not* be a member of a party during his actual term.[61]

Changes and corrections in late 1997 were introduced into the 'Law about the election of the President of the Republic of Uzbekistan', in which a candidate standing for president no longer needed to be younger than 65 but did have to be proficient in the Uzbek language and have permanently lived on the territory of Uzbekistan for at least the previous ten years. In addition, the two-term limit clause was lifted from the presidency.[62] This not only allowed Karimov the opportunity to become a de facto president for life, it also constrained his only potential rivals: they tended to live outside of Uzbekistan and criticized his administration from abroad. Under the new law they were no longer eligible to run for president as they violated the permanent resident clause.

In 2000 the presidential elections were 'contested' by Karimov and the first secretary of the NDPU, Karimov's former party and still his personal political fiefdom. The other parties officially 'registered' and allowed to participate in the elections actually *nominated* Karimov. In other words, out of five official parties four nominated Karimov and the fifth put forth a candidate who declared on election-day that *he himself was voting for the incumbent*.

Another referendum put forward in January 2002 reformed the parliament from a unicameral to a bicameral body and formally extended the term of president from five to seven years. While the international community saw the presidential term switch for what it was (a brazen power extension that effectively gave Karimov the theoretical opportunity to be in power for 23 consecutive years), the bicameral parliamentary reform was hoped to be an improvement. This hope was misplaced. The lower parliament, the National Assembly, consisted of 120 members. Only registered parties could field candidates for elections. In Uzbekistan there were only five formally registered parties with all of them being pro-presidential to one degree or another. As a consequence almost all of the deputies were pro-Karimov. The upper chamber, the Senate, consisted of 100 members – 16 directly appointed by the president and the remaining 86 elected at sessions of district, regional and city deputies. Local municipalities across Uzbekistan were presidential designations beholden to the president. Thus the entire Senate became a de facto presidential appointment.[63]

The first bicameral parliamentary elections, contested in late 2004, were a farce. Not one individual opposition candidate stood in the 2004 elections. Central Electoral Commission officials outright refused to register opposition candidates or invoked insanely confusing registration procedures that ensured the candidates' failure to comply. As a result no opposition movement was successfully legalized before the elections.[64] More than 500 candidates competed for the 120 seats but all of them were from pro-government parties which had little to no divergence in party platforms from the executive administration. In fact, Karimov had succeeded in turning the entire parliament into a rich boy's club as most of the candidates were businessmen with close ties to the president.[65] Most international watch-dog groups did not bother to send monitors as it was a foregone conclusion that falsification and fraud would be widespread.

The free expression of the media seemed to be judged by the international community as a success if it simply could succeed in not getting worse in Uzbekistan, but get worse it did. In May 2002 the Uzbek government officially banned pre-publication censorship. This was considered a major step forward by Uzbekistan but in reality simply meant pre-publication censorship was replaced by a thorough and intensive post-publication review.[66] In fact, in some ways the law only *increased* censorship as editors, ever fearful of government retribution, cracked down on their own papers making sure no potentially offending pieces made it to press. Indeed, this was the exact purpose in enacting the reform. As such it could claim compliance with international norms of freedom of expression while still ensuring media servitude. As one leading Uzbek journalist explained:

> Information about the activities of the government and the president is divided into the internal and the external. For society, official external doses of information are given out, minimal in the extreme, a long way from the truth, and at times simply laughable in their absurdity. And what they do in reality, what their aims are – that remains a big secret.[67]

The reason self-censorship was so pervasive was not simply because of fear of repression, though surely that existed. Most media remained financially under state control. Thus even if editors were not fearful of expressing their political views they undoubtedly had fears of the bottom line: a simple phone call from the executive administration resulted in the immediate removal of state financing which would result in a total shutdown. The combination of business and personal survival was a powerful mix constraining journalistic professionalism.

The 'Uzbek way' for economic development was meant to be a localized version of the China program – strict state control with a gradual loosening of the reigns over time, only making sure a significant percentage of the profits continued going to government coffers. The real Uzbek model played out as a sycophantic system of client-patron nepotism. Very little true privatization took place in Uzbekistan, with the vast majority of industries remaining in government hands with a controlling interest or going to cronies loyal to the administration.[68] As a consequence the government had no real sense of economic reality in the country.

> Senior officials are afraid to admit honestly to the president that there is a negative tendency in some sectors of the economy…government officials prefer to compromise their honesty rather than lose the power and wealth they get in return for loyalty.[69]

This resulted in business being conducted only through connections. It ran the gamut from industry to trade to banking. No aspect of the Uzbek economy was immune. This reliance on personal connections mutated over time into shadow economies as people without the right acquaintances did what they must to either obtain or work without them. Thus corruption became a block deterring any sincere attempt at reform or legitimate economic participation.

In theory a true market economy gradually co-opts economic elites because they come to see the natural advantages in entering the market – their immediate dominance combined with the open environment allows for massive profits to be made that could not be obtained under the state-command nepotism model. But in a system like Uzbekistan, where economic and political elites were fused beyond distinction, this co-optation was hard to achieve as elites had no experience in engaging legitimate and fair competition. In addition, an open competition across the economy would inevitably spread access and influence across the political sphere where they up to then had unfettered control.[70] As such there was not only pressure on present elites to maintain the status quo, there was also tremendous pressure on up and coming elites to simply follow in their footsteps. In Uzbekistan, the young generation that so often was the true hope of the international community to make real change only saw three realistic choices:

- Go corrupt and preserve the system
- Attempt to change the system within, be dominated and possibly die
- Rebel against the system, go underground and become radical.

There is little hope that Uzbekistan can change gradually from within. The next generation has been in essence co-opted before they even leave grade school.

There was also hope that change could be fostered through non-governmental organizations (NGOs) and civil society activism. This was also a myth. In theory, the presence and expansion of such groups only increases economic and political openness.[71] In reality, Karimov added legislation that severely restricted their freedom of assembly. In January 2003 a presidential decree restricted gatherings of more than 100 people to organizations with special permission only.[72] Of course that special permission was nearly impossible to obtain. In December 2003 the National Security Service (NSS) established a separate department in order to monitor and control NGOs, specifically their contacts with international organizations and what they revealed in candid interviews to foreign media.[73] The Tax Inspectorate also joined in the fray by making banks obtain permission from it before being able to release grant money to particular NGOs.

Finally, the Ministry of Foreign Affairs announced in December 2003 that all foreign NGOs were obligated to re-register with the Ministry of Justice by March 2004. The government claimed this was merely a technicality but it was clear the maneuver was used as a club to limit if not outright shut down foreign NGOs meddling in Uzbek affairs.[74] Continuing violence up to and through 2005 made Karimov's paranoia toward NGOs quite broad: he aggressively harassed, intimidated and 'encouraged' NGOs, in the Ferghana Valley in particular, to 'apply for self-liquidation'.[75] According to some estimates as many as 1,600 local NGOs in the Ferghana Valley alone were forced to close operations in 2005.[76] Reasons or official explanations for the closings were never given. As the deputy chief of one NGO, based in Andijan, said in late 2005:

> I was summoned to the local hokimiyat (municipality) and told to apply for self-liquidation. When I asked why they said there was an order from high up. I asked if I could see the order, but they said, apply for closure in a friendly way, or we will close you down in an unfriendly way.[77]

In 2005 Andijan was the scene of a bloody protest/massacre that brought to a head all the contradictions and hypocrisies of the Karimov regime. Unfortunately, how the massacre has played out since only further exposes an international community inexplicably lacking the political will to do the right thing. As Andijan and subsequent changes in Uzbek foreign policy is discussed, it will become obvious that American 'engagement with principles' achieved nothing except the US first being used and then discarded when it was no longer expedient to Karimov.

The Andijan Massacre

Out of all the various reports coming out since the Andijan incident, by far the most complete was done by Human Rights Watch, which undertook to discover the smaller sparks that led to the huge civil explosion and government retaliation. Excerpts from the lengthy report are highlighted below:[78]

The trigger for the Andijan protests was the June 2004 arrest of twenty-three successful local businessmen on charges of religious extremism, for their alleged membership in a banned Islamic movement, Akramia. Some observers saw the prosecution as a reaction to the businessmen's growing authority in the Andijan community, garnered from having provided relatively high wages and good benefits to their employees. As the trial progressed from February 2005 into May, the businessmen's supporters began to protest the hearings. Popular discontent grew and, on May 10, some seven hundred to one thousand people gathered outside the Altinkul District Court to protest the proceedings…

As Andijan awaited a verdict in the trials on May 12, relatives and supporters of the businessmen took action. Around midnight on May 12 a group of between fifty and one hundred men attacked a local police station and then stormed the Ministry of Defense barracks No.34, seizing weapons and a military vehicle. The armed group then broke through the gates of the Andijan prison where the twenty-three businessmen were held. They freed the businessmen and hundreds of additional inmates [also held as religious extremists]. The men then moved to take control of the hokimat (local municipality building), with some of the group engaging in a heavy gun battle with security officials outside the National Security Service headquarters.

As the crowd grew on Bobur Square, the gunmen started taking law enforcement and government officials as hostages…As the day went on, Uzbek security forces indiscriminately shot into the crowd from armored personnel carriers and sniper positions above the square. Towards the evening government troops blocked off the square and then, without warning, opened fire, killing and wounding unarmed civilians. People fled the square in several groups…As they tried to escape hundreds of people were shot by snipers or mowed down by troops firing from the APCs. After the peak of the carnage, government forces swept through the area and executed some of the wounded where they lay…

In the immediate aftermath of the massacre government authorities closed off Bobur Square and Cholpon Prospect where much of the killing had taken place. The bodies were removed and evidence of the massacre was erased. Authorities washed the blood from the street and painted over the bullet-riddled buildings of the surrounding neighborhood. The government stationed armed guards around the local hospitals and forbade independent journalists and human rights investigators access to the hospitals, morgues and cemeteries. Foreign journalists were detained by police, threatened and forcibly evicted from the city…In the hours and days that followed Andijan became a closed city…

The government of Uzbekistan has characterized the killings in Andijan as terrorist acts and put the death toll at 187, the majority of them bandits, terrorists, and the government agents they supposedly killed. It has acknowledged sixty civilian deaths and has attributed all of them to the gunmen

> and not to fire by government forces. The government has specifically claimed that the gunmen were the ones responsible for the slaughter of civilians retreating from the main square where the protest had been held. Government officials have stated publicly that foreign powers, a barely veiled reference to Western governments, instigated the uprising with the aim of carrying out revolts in Uzbekistan similar to those in Georgia, Ukraine and Kyrgyzstan…
>
> The government and state-controlled media in Uzbekistan are working furiously to rewrite history, to produce a new account of the Andijan events, and to bury the facts that contradict it.

The international community, namely the European Union and the United Nations, called for an international inquiry and threatened that strong measures would be taken if Karimov did not cooperate by the end of June 2005.[79] Cooperation did not come. Neither, unfortunately, did any strong measures. For whatever reason, the international community continued to believe in a fantasy that Karimov would respond to soft pressure. In reality, over the course of fifteen years Karimov has continuously betrayed the international community and its conventions. Andijan proved to be no different.

By mid-November justice authorities had brought to trial 15 defendants from the Andijan incident. None of them belonged to government forces. In what was clearly a show trial that violated every possible international covenant on fair-trial standards, the Uzbek Supreme Court handed down a guilty verdict on over thirty counts, including membership in an extremist organization, murder and terrorism.[80] Sentences ranged from fourteen to twenty years in prison.

In what was no doubt a testament to the continued systematic use of torture, the trial actually began with all of the defendants confessing to the charges, with several of them *even requesting to be given the death penalty*.[81] It seems likely that the defendants made such declarations in return for family being left alone. Their own testimony, instead of being an attempt to present their side of the story, closely followed the official prosecutorial indictment. In fact, some of the defendants' testimony actually copied the indictment verbatim.[82] Six defense lawyers actually began opening remarks by *apologizing and begging forgiveness* for defending such 'guilty persons'.[83] Perhaps most disturbing of all, these fifteen defendants did not represent any sort of symbolic atonement for the overuse of government power. Officially more than one hundred people are still waiting in prisons, waiting to be brought to trial on the same charges and under the same maniacal fraud. No high-level government officials have been even charged.

Andijan shows what happens when the international community chooses appeasement over enforcement and then rationalizes its own position. This was the case with the US when the state department gave its strongest condemnation of the violence by simply saying it was 'deeply disturbed' by reports that unarmed demonstrators had been fired on. At the same time State Department spokesman Richard Boucher also condemned the violent protesters who had stormed government buildings, parroting Karimov's concerns about Islamic extremists.[84] The reason for the storming, as elaborated above, was conveniently bypassed.

Even more unfathomable was the continued belief by some experts to view the oppression as an opportunity rather than a problem. Chris Seiple, head of the Institute for Global Engagement, was a perfect example of this unfounded hopefulness:

> This crisis can bring a new phase to our relationship. We can say to President Karimov, you have a choice: work with us and engage in reform, or we will not be there in the long term. The benefits of our base in Uzbekistan are tactically important but strategically insignificant.

Perhaps the only thing equal to Karimov's audacity for repression was the West's own arrogance and misplaced belief that it operated from a position of strength with Uzbekistan. Seiple's comment simply stated it outright: Karimov should follow Western interests because the potential consequences of not having the US as a friend would be too dire. Karimov apparently didn't see it that way. If he saw the color revolutions across the region as anything he saw them as business as usual, the replacement of one elite group by another and then cloaked in the guise of democracy. If he saw the United States as anything he saw it as a tenant. As everyone knows, the biggest concern for a tenant should be avoiding eviction. Karimov finally tired of an 'ally' that was quickly becoming only a toothless nuisance with its hypocritical criticism. We did not give him fear, only irritation. Thus he chose to evict the US from its airbase, beginning in January 2006, and then went on to sign an old-school mutual non-aggression pact with the Russia.

It is doubtful that so-called experts envisioned this as the 'new phase' in the US-Uzbek relationship after Andijan. In short, the West got played and played very badly. Unfortunately, the losers in this game are legion: in the short-term, Uzbek society is devastated. In the long-term, the US comes under greater threat for constantly allowing itself to be seen across the Ferghana Valley as the tacit support that allowed torture and gave Karimov the implicit justification for the repression of democratic rights and a systematic program of human suffering.

Conclusion – 'A Cloud-Cuckoo-Land Kind of Place'

Reform through positive engagement was an illusion simply because our engagement didn't demand any real reform. We watched complacently as Uzbek politics and society degraded until it bordered on the surreal. As former British Ambassador to Uzbekistan Craig Murray argued recently:

> [Uzbekistan] is a kind of cloud-cuckoo-land place…It's a completely mad totalitarian society. The government lies all the time…Karimov's politics are essentially paranoid. He has a paranoid view of the world…who's collected hundreds of millions of dollars, stolen from the people. [Torture] is completely

systematic and not rare at all…in both political and criminal cases I would guess in over 95% of cases the accused person signs a full confession.

The Americans amazed me with their hypocrisy [when Condoleeza Rice named Cuba, Iran, North Korea, Zimbabwe, Burma and Belarus as outposts of tyranny, and left out Uzbekistan]…I saw a most wonderful statement from the American ambassador, a load of pious rubbish, where he applauded the elections as a step on the road to democracy and then at the end of it said it was unfortunate that the opposition weren't allowed to take part!...Uzbekistan is, by any measure, a much worse dictatorship than [many of the countries on the Secretary's list] and Condoleeza Rice is just talking, well, crap.[85]

Map 6.1 Uzbekistan in Central Asia

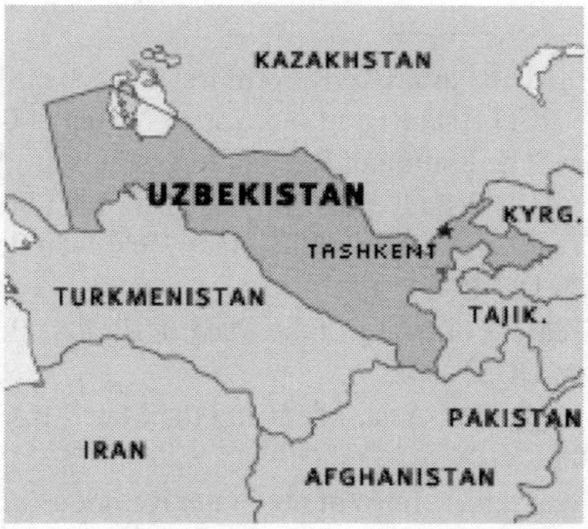

Colorful language aside, Ambassador Murray's assessments are accurate. Already across Uzbekistan differences are readily apparent in the general societal attitudes toward radical Islam. More disturbing, it is also clear that those sympathies are simultaneously converging with disdain for the United States.[86]

The noticeable change is that people have become more religious. The people that I knew three years ago, that didn't want to even hear about Islam, now walk around with little prayer books in their hands.

Before I was afraid to say anything to you but now on the contrary I want to let everybody know what is really going on in Uzbekistan. I would rather live under Islamists than under these democrats.

The [Americans] provide the guns to the police and the cars that they take us away in and the hundreds of millions of dollars to the government that means they don't do any reforms and I don't understand it.

> I had heard a lot about American democracy. I thought that the appearance of American troops here would change the situation for the better. Now I see that the regime has only been strengthened and arrests and abuses only increased.

Many analysts in human rights organizations and grass-roots movements have long argued that the fundamental short-sightedness of the Karimov regime has made Uzbekistan, and particularly the Ferghana Valley part of its country, fertile ground for future Islamic terrorists. What they miss and this work wants to emphasize is that while Karimov may have planted the seeds of Islamic hatred himself it was the US acting as enabler that provided the fertilizer. It is at our own mortal risk if we continue to be naïve enough to think the fertilizer we provided is sufficient only for growing weeds that can choke Uzbekistan.

---

1 International Crisis Group, 'The Failure of Reform in Uzbekistan: Ways forward for the International Community', ICG Asia Report No. 76, 11 March 2004.

2 International Crisis Group, 'The Failure of Reform in Uzbekistan: Ways forward for the International Community', ICG Asia Report No. 76, 11 March 2004.

3 Human Rights Watch, 'Creating Enemies of the State: Religious Persecution in Uzbekistan', New York and Washington DC, 2004.

4 Human Rights Watch, 'Creating Enemies of the State: Religious Persecution in Uzbekistan', New York and Washington DC, 2004.

5 International Crisis Group, 'Central Asia: Islam and the State', ICG Asia Report No. 59, Osh/Brussles, 10 July 2003.

6 Human Rights Watch, 'Creating Enemies of the State: Religious Persecution in Uzbekistan', New York and Washington DC, 2004.

7 International Crisis Group, 'Central Asia: Islam and the State', ICG Asia Report No. 59, Osh/Brussles, 10 July 2003.

8 Igor Rotar, 'Enlightened Islam, Uzbek-style: Islam Karimov is getting rid of his most dangerous rivals', *Prism*, Volume 4 Issue 13, Washington DC, 26 June 1998.

9 Igor Rotar, 'Enlightened Islam, Uzbek-style: Islam Karimov is getting rid of his most dangerous rivals', *Prism*, Volume 4 Issue 13, Washington DC, 26 June 1998.

10 Igor Rotar, 'Enlightened Islam, Uzbek-style: Islam Karimov is getting rid of his most dangerous rivals', *Prism*, Volume 4 Issue 13, Washington DC, 26 June 1998.

11 Human Rights Watch, 'Creating Enemies of the State: Religious Persecution in Uzbekistan', New York and Washington DC, 2004.

12 Human Rights Watch, 'Creating Enemies of the State: Religious Persecution in Uzbekistan', New York and Washington DC, 2004.

13 International Crisis Group, 'Central Asia: Islam and the State', ICG Asia Report No. 59, Osh/Brussles, 10 July 2003.

14 Human Rights Watch, 'Creating Enemies of the State: Religious Persecution in Uzbekistan', New York and Washington DC, 2004.

15 International Crisis Group, 'Central Asia: Islam and the State', ICG Asia Report No. 59, Osh/Brussles, 10 July 2003.

16 International Crisis Group, 'Central Asia: Islam and the State', ICG Asia Report No. 59, Osh/Brussles, 10 July 2003.

17 International Crisis Group, 'Central Asia: Islam and the State', ICG Asia Report No. 59, Osh/Brussles, 10 July 2003.

18 International Crisis Group, 'Central Asia: Islam and the State', ICG Asia Report No. 59, Osh/Brussles, 10 July 2003.

19 International Crisis Group, 'Central Asia: Islam and the State', ICG Asia Report No. 59, Osh/Brussles, 10 July 2003.

20 International Crisis Group, 'Central Asia: Islam and the State', ICG Asia Report No. 59, Osh/Brussles, 10 July 2003.

21 Scott Peterson, 'Amid Reforms, Muslims Still Under Fire in Ferghana Valley', Christian Science Monitor, 19 July 2002.

22 Human Rights Watch, 'Creating Enemies of the State: Religious Persecution in Uzbekistan', New York and Washington DC, 2004.

23 Human Rights Watch, 'Creating Enemies of the State: Religious Persecution in Uzbekistan', New York and Washington DC, 2004.

24 Human Rights Watch, 'Creating Enemies of the State: Religious Persecution in Uzbekistan', New York and Washington DC, 2004.

25 Human Rights Watch, 'Creating Enemies of the State: Religious Persecution in Uzbekistan', New York and Washington DC, 2004.

26 Human Rights Watch, 'Creating Enemies of the State: Religious Persecution in Uzbekistan', New York and Washington DC, 2004.

27 Human Rights Watch, 'Creating Enemies of the State: Religious Persecution in Uzbekistan', New York and Washington DC, 2004.

28 Human Rights Watch, 'Creating Enemies of the State: Religious Persecution in Uzbekistan', New York and Washington DC, 2004.

29 Human Rights Watch, 'Creating Enemies of the State: Religious Persecution in Uzbekistan', New York and Washington DC, 2004.

30 All article elaborations are taken from Human Rights Watch, 'Creating Enemies of the State: Religious Persecution in Uzbekistan', New York and Washington DC, 2004.

31 Human Rights Watch, 'Creating Enemies of the State: Religious Persecution in Uzbekistan', New York and Washington DC, 2004.

32 Human Rights Watch, 'Creating Enemies of the State: Religious Persecution in Uzbekistan', New York and Washington DC, 2004.

33 Human Rights Watch, 'Creating Enemies of the State: Religious Persecution in Uzbekistan', New York and Washington DC, 2004.

34 Human Rights Watch, 'Creating Enemies of the State: Religious Persecution in Uzbekistan', New York and Washington DC, 2004.

35 Human Rights Watch, 'Creating Enemies of the State: Religious Persecution in Uzbekistan', New York and Washington DC, 2004.

36 Human Rights Watch, 'Creating Enemies of the State: Religious Persecution in Uzbekistan', New York and Washington DC, 2004.

37 Human Rights Watch, 'Creating Enemies of the State: Religious Persecution in Uzbekistan', New York and Washington DC, 2004.

38 Human Rights Watch, 'Creating Enemies of the State: Religious Persecution in Uzbekistan', New York and Washington DC, 2004.

---

[39] Human Rights Watch, 'Creating Enemies of the State: Religious Persecution in Uzbekistan', New York and Washington DC, 2004.

[40] Human Rights Watch, 'Creating Enemies of the State: Religious Persecution in Uzbekistan', New York and Washington DC, 2004.

[41] Human Rights Watch, 'Creating Enemies of the State: Religious Persecution in Uzbekistan', New York and Washington DC, 2004.

[42] Human Rights Watch, 'Creating Enemies of the State: Religious Persecution in Uzbekistan', New York and Washington DC, 2004.

[43] Human Rights Watch, 'Creating Enemies of the State: Religious Persecution in Uzbekistan', New York and Washington DC, 2004.

[44] Aljazeera New Service, 'Bereaved Uzbek Mother Gets Prison', http://english.aljazeera.net/NR/exeres/9B813F38-57CC-4DF0-8147-2D3F38B04316.htm, accessed 21 December 2004.

[45] Department of Defense, 'Secretary of Defense Rumsfeld and Ambassador Purnell Press Conference in Uzbekistan', http://www.defenselink.mil/transcripts/2004/tr20040225-secdef0492.html, accessed 23 December 2004.

[46] Jim Lobe, Muslim Uzbekistan, 'With Friends Like Uzbekistan', http://archive.muslimuzbekistan.com/eng/ennews/2004/04/ennews01042004b.htm, accessed 14 November 2005.

[47] Department of Defense, 'Secretary Rumsfeld Press Conference with President of Uzbekistan', http://www.defenselink.mil/transcripts/2001/t10082001_t1005uz.html, accessed 4 January 2003.

[48] All excerpts taken from US State Department, 'Joint Press Conference with Treasury Secretary O'Neill and President of Uzbekistan Karimov', http://www.state.gov/e/eb/rls/rm/2002/11945.htm, accessed 8 March 2004.

[49] All excerpts taken from Department of Defense, 'Secretary Rumsfeld and Ambassador Jon Purnell Press Conference in Uzbekistan', http://www.defenselink.mil/transcripts/2004/tr20040225-secdef0492.html, accessed 18 June 2005.

[50] Ariel Cohen, 'After the Uzbekistan Terror Attacks: Don't Sanction, but Press for Democracy', The Heritage Foundation, http://www.heritage.org/Research/RussiaandEurasia/wm496.cfm, accessed 12 October 2004.

[51] Ariel Cohen, 'After the Uzbekistan Terror Attacks: Don't Sanction, but Press for Democracy', The Heritage Foundation, http://www.heritage.org/Research/RussiaandEurasia/wm496.cfm, accessed 12 October 2004.

[52] Ariel Cohen, 'After the Uzbekistan Terror Attacks: Don't Sanction, but Press for Democracy', The Heritage Foundation, http://www.heritage.org/Research/RussiaandEurasia/wm496.cfm, accessed 12 October 2004.

[53] Ariel Cohen, 'After the Uzbekistan Terror Attacks: Don't Sanction, but Press for Democracy', The Heritage Foundation, http://www.heritage.org/Research/RussiaandEurasia/wm496.cfm, accessed 12 October 2004.

[54] The Bureau of European and Eurasian Affairs, 'US Government Assistance to and Cooperative Activities with Eurasia – 2004 Report', Washington DC, January 2004.
[55] The Bureau of European and Eurasian Affairs, 'US Government Assistance to and Cooperative Activities with Eurasia – 2004 Report', Washington DC, January 2004.
[56] The Bureau of European and Eurasian Affairs, 'US Government Assistance to and Cooperative Activities with Eurasia – 2004 Report', Washington DC, January 2004.
[57] The Bureau of European and Eurasian Affairs, 'US Government Assistance to and Cooperative Activities with Eurasia – 2004 Report', Washington DC, January 2004.
[58] The Bureau of European and Eurasian Affairs, 'US Government Assistance to and Cooperative Activities with Eurasia – 2004 Report', Washington DC, January 2004.
[59] The Bureau of European and Eurasian Affairs, 'US Government Assistance to and Cooperative Activities with Eurasia – 2004 Report', Washington DC, January 2004.
[60] Eurasianet, 'Presidential Elections in Uzbekistan in 2000', http://www.eurasianet.org/departments/election/uzbekistan/presanal.htm, accessed 12 July 2003.
[61] Eurasianet, 'Presidential Elections in Uzbekistan in 2000', http://www.eurasianet.org/departments/election/uzbekistan/presanal.htm, accessed 12 July 2003.
[62] Eurasianet, 'Presidential Elections in Uzbekistan in 2000', http://www.eurasianet.org/departments/election/uzbekistan/presanal.htm, accessed 12 July 2003.
[63] Electionworld, 'Elections in Uzbekistan', http://www.electionworld.org/uzbekistan.htm, accessed 4 September 2005.
[64] Eurasianet, 'Uzbekistan Holds Elections – Without Opposition', http://www.eurasianet.org/departments/civilsociety/articles/eav12230.htm, accessed 16 September 2004.
[65] Eurasianet, 'Uzbekistan Holds Elections – Without Opposition', http://www.eurasianet.org/departments/civilsociety/articles/eav12230.htm, accessed 16 September 2004.
[66] International Crisis Group, 'Uzbekistan's Reform Program: Illusion or Reality?' Asia Report No. 46, Osh/Brussels, 18 February 2003.
[67] International Crisis Group, 'Uzbekistan's Reform Program: Illusion or Reality?' Asia Report No. 46, Osh/Brussels, 18 February 2003.
[68] International Crisis Group, 'Uzbekistan's Reform Program: Illusion or Reality?' Asia Report No. 46, Osh/Brussels, 18 February 2003.
[69] International Crisis Group, 'Uzbekistan's Reform Program: Illusion or Reality?' Asia Report No. 46, Osh/Brussels, 18 February 2003.
[70] International Crisis Group, 'Uzbekistan's Reform Program: Illusion or Reality?' Asia Report No. 46, Osh/Brussels, 18 February 2003.
[71] International Crisis Group, 'The Failure of Reform in Uzbekistan: Ways Forward for the International Community', Asia Report No. 76, Osh/Brussels, 11 March 2004.
[72] International Crisis Group, 'The Failure of Reform in Uzbekistan: Ways Forward for the International Community', Asia Report No. 76, Osh/Brussels, 11 March 2004.
[73] International Crisis Group, 'The Failure of Reform in Uzbekistan: Ways Forward for the International Community', Asia Report No. 76, Osh/Brussels, 11 March 2004.

[74] International Crisis Group, 'The Failure of Reform in Uzbekistan: Ways Forward for the International Community', Asia Report No. 76, Osh/Brussels, 11 March 2004.

[75] Eurasianet, 'Uzbek Government Exerting Pressure on Local NGOs to Close Voluntarily', http://www.eurasianet.org/departments/civilsociety/articles/eav100405.shtml, accessed 13 November 2005.

[76] Eurasianet, 'Uzbek Government Exerting Pressure on Local NGOs to Close Voluntarily', http://www.eurasianet.org/departments/civilsociety/articles/eav100405.shtml, accessed 13 November 2005.

[77] Eurasianet, 'Uzbek Government Exerting Pressure on Local NGOs to Close Voluntarily', http://www.eurasianet.org/departments/civilsociety/articles/eav100405.shtml, accessed 13 November 2005.

[78] Human Rights Watch, 'Burying the Truth: Uzbekistan Rewrites the Story of the Andijan Massacre', Vol. 17, No. 6(D), September 2005.

[79] Steve Crawshaw, 'Getting Away with Murder', *The Guardian*, 31 August 2005.

[80] Human Rights Watch, 'Uzbekistan: Andijan Show Trial Ends with Guilty Verdict', http://hrw.org/english/docs/2005/11/14/uzbeki12020.htm, accessed 23 November 2005.

[81] Human Rights Watch, 'Uzbekistan: Andijan Show Trial Ends with Guilty Verdict', http://hrw.org/english/docs/2005/11/14/uzbeki12020.htm, accessed 23 November 2005.

[82] Human Rights Watch, 'Uzbekistan: Andijan Show Trial Ends with Guilty Verdict', http://hrw.org/english/docs/2005/11/14/uzbeki12020.htm, accessed 23 November 2005.

[83] Human Rights Watch, 'Uzbekistan: Andijan Show Trial Ends with Guilty Verdict', http://hrw.org/english/docs/2005/11/14/uzbeki12020.htm, accessed 23 November 2005.

[84] Matthew Davis, 'Uzbek Crisis Poses Dilemma for US', BBC News, 16 May 2005.

[85] Interview with Craig Murray, Turkishweekly, 24 January 2005, http://www.turkishweekly.net/interview.php?id=48, accessed 5 May 2005.

[86] All of the proceeding quotes are taken from interviews conducted by International Crisis Group in Tashkent, 'The Failure of Reform in Uzbekistan: Ways Forward for the International Community', Asia Report No. 76, Osh/Brussels, 11 March 2004.

Chapter 7

Inside the Cauldron of the Hizb ut-Tahrir

The present work has focused on how the permissive willingness of the United States to overlook and tacitly accept horrendous domestic repression against democracy and Islam within the Ferghana Valley states (when such willingness is a contradiction of US policy, a betrayal of our long-term security interests and possibly even a legal violation of US law) was likely to cause an increase in radical Islamist groups. In addition, these groups would not be committed to focusing exclusively on their domestic oppressors but were also likely to take aim at targets beyond Central Asia. In essence, American foreign policy has helped the valley become a haven for the exact kind of fundamentalist Islam that the US declared war against in 2001.

There has clearly been a boom of radical groups within the valley. Tabligh (Mission), Uzun Sokol (Long Beard), Adolat Uyushmasi (Justice Society), Islam Lashkarlari (Warriors of Islam), Tovba (Repentance), and Nur (Ray of Light) is but a small sampling slowly making headway within the region. For the most part these groups are small and isolated and do not pose an immediate threat to Uzbekistan, Kyrgyzstan and Tajikistan. This, however, is false relief, as these groups are already open believers in the 'clash of civilizations' that Osama bin Laden fervently supports. This work is not so much concerned with the military capabilities of radical Islam within the Ferghana Valley. It is concerned with how strongly radical groups empathize with bin Laden's philosophy and how they may indirectly and directly become a part of and further his cause.

Of all the groups operating within the valley it is the Hizb ut-Tahrir (HT) that earns exclusive focus in this chapter. An international movement long stationed in Ferghana (and long oppressed by the local regimes), HT (The Party of Liberation) is easily the most famous fundamentalist Islamic organization presently in Central Asia. This chapter studies the actual beliefs and ideology of HT. It reveals that while the HT hierarchy publicly proclaims peaceful revolution its actual propaganda is likely to produce sympathizers and admirers who are not so committed.[1] More importantly, it is clear that HT's ideology is obsessed not with just the Ferghana regimes but with the United States and considers it as the primary target of its derision and animosity.

A Little Historical Background

HT grew out of various movements in the Middle East in the 1950s but is considered to have been formally founded by Sheikh Taquiuddin an-Nabhani al Falastini in Jordanian-controlled East Jerusalem.[2] It has been present in Central

Asia at least since the early 1990s but made more dramatic headway with local populations later on in the decade as independence began to transform into repression. HT is a legitimate transnational organization with considerable support across Western Europe among young Muslims and an expansive organizational base in London.[3] The extent of its reach and details about its membership remain murky at best but its numbers within Central Asia certainly go into the thousands.

HT's claim to fame up to now has been its call for the reestablishment of an Islamic Caliphate that would unite all Muslims regardless of national, regional, tribal or clan differences. Its propaganda is vehement in its denunciation of the West and rejection of Arab regimes not properly honoring Islamic heritage. It claims to support only non-violent methods but is not against revolutionary struggles conducted by other groups that already exist and do involve violence.[4] In many ways HT mimics the 'non-involvement' of Osama bin Laden, who personally does not take part in violent operations but clearly adheres to an ideology and belief-system that implicitly accepts, applauds and aids them. Thus, HT's so-called pacifism needs to be understood as a false idol: its non-violence pledge is most certainly countered by the tacit support it gives of others' violence.

HT ideology is a marriage of Islamist theology with Leninist-Marxist beliefs.[5] As a consequence its criticisms have been too easily dismissed by Western governments as totalitarian theocratic hogwash. I believe this dismissal to be incorrect and dangerous. While their political criticisms may be unfamiliar to Americans, many do accurately critique the kind of democracy being built in Central Asia. Consequently, what makes the HT look like a fringe group to the West is also likely give it great legitimacy with the people in the Ferghana Valley.

Democracy – The Faithless System[6]

> Democracy, brought by Western infidels to Islamic countries, has no relation whatsoever to Islam. It is a system which, by its general conceptions, details, evolution, base principles and foundation, is wholly antithetical to Islam. For this reason it is forbidden to any Muslim to accept or participate in a democracy.
>
> Democracy is a system created by people who sought to take refuge from… leadership in the name of religion. Thus the foundations of this system are people who do not wish to have any relation to heaven and religion in general.

This is not in fact blatant misinformation. HT accurately goes on to describe the evolution of democracy as being a response to the injustice of European monarchs who sat upon a throne justified by their so-called anointment from God. But what most western analysts miss is the small subtlety which causes HT to dismiss this evolution as a negative trend for human government.

> The word democracy came from the west, meaning self-government of the people by their own laws. This means that the people are the unfettered holders of power. They hold hegemony over power - it can belong to no other body or

authority. The people are the definers of their own will…The people are not responsible before any other power besides their own…

There are four principles of freedom which express the ability of the people to exercise full power without any form of barrier blocking their hegemony:

- Religious freedom
- The freedom of expression
- The freedom of private property
- Personal freedom.

All of these freedoms, even the very idea of democracy itself, emerged from a secularism that was embedded within capitalist economic philosophy. From this secularism naturally emerged the concept of separating religion from the state. This idea became a foundational doctrine…This idea would eventually become so important that it can be argued that it is the ideological foundation of the West itself, making all of its citizenry beholden to the principle…This doctrine isolates religion and the church from life, from the state, as well as from the process of adopting laws and supporting governmental authority…From this the following can be concluded:

- Democracy does not maintain any godly foundation. It is a product purely of the human imagination. It does not have any divine inspiration nor does it seek any connection to the wisdom of the prophets of God.
- Democracy's secularism is founded on two ideas – that the people hold hegemony and that the people are the base for all power.

The will of the majority is not a great thing because it is still in the end founded on sinful and imperfect humanity. The very phrase that causes most admirers of Western democracy to swell with pride causes HT great consternation - the hegemony of the people over the rule of law. If the foundation is flawed, then the entire system must be as well.

The separation of religion from state is a negative because it severs the only real connection humanity has with morality. In fundamentalist Islam there is a deep-seated belief in the innate badness of humans (which is not unfamiliar to Christians who respect the concept of original sin). In fact it is Adam's fall from grace, his conscious decision and exercise of free will, which led to his banishment and the subsequent downfall of all humanity. Creating a system, therefore, that *purposefully* tries to legally disconnect the state away from religion and which strives to remove religion from decision-making power is a system that elevates humanity to the level of God. Assuming democracy plays out in reality exactly as it is written on paper this is still wrong to HT on the theoretical level. HT, however, does not grant that assumption. When the empirical reality of democracy around the globe is taken into account HT only becomes more vehemently anti-democratic and against its inherent freedoms.

> Before anyone analyzes the antithetical nature of democracy to Islam, it should first be noted that true democracy is empirically absent in the real world, even in those countries where it has been established the longest. This is because democracy has been established everywhere on a base of lies and deceit, which has brought nothing but poverty and suffering to the entire world…
>
> Democracy by its very essence is a utopian idea which is not destined to truly arise, not now or in the future. Though the idea of direct democracy is accepted as impossible, even the subsequent alternatives (introducing the institutions of president, parliament and government, for example) do not come close in reality to what they are meant to be theoretically…
>
> In fact, democracy has never truly been realized anywhere in the world. The members of parliament are not elected by the majority of voices but by the minority as the very process of competition splits votes and creates a victor with minimal overall popular support…It should be remembered that the two oldest democracies in the world, the United States and Great Britain, elect heads of states and members of parliament mainly by the will of the capitalists, especially via the major business leaders and industrial monopolies. They in no way represent the will of the people or even the will of the majority. These major capitalists support financially only those candidates who can be guaranteed to act in the name of their interests, not the interests of the general people…Thus it can be said that democracy is not the will of the people, but the will of capitalism…they do not receive power in the name of the people, they take their power from the people.

Considering the most recent controversies in presidential elections in the United States (and the Capitol Hill corruption scandal plodding along as I write), it is difficult to objectively say there is no basis for HT's arguments. More importantly to this study, however, is to view HT's analysis of democracy in light of the 'democratic' regimes in the Ferghana Valley states. These democracies *are* elite-dominated criminocracies that serve only the interests of the few. In fact, such accusations are actually a *kind* portrayal of the Uzbek, Tajik and Kyrgyz regimes.

Taken at face value too much of the HT ideology can read like a starter set for Islamic Marxism. In the post-Cold War world that too often leads to early dismissal by Western scholars. But in places like the Ferghana Valley the criticisms embedded within Marxism remain incredibly pertinent. The discussion of corporate monopolies that act in full cooperation with corrupt and powerful governments, preying on societies unable to defend themselves militarily nor compete economically, rings true in the valley. As a consequence HT reveals a disturbing and bitter truth to common Muslims: democracy can continue without them. Their lives are irrelevant to the world system. Their own government leaders have prostituted themselves and their countries resources only for personal gain and comfort. The cloak of quasi-Marxism may act an intellectual barrier to Western diplomats but it is not to common Muslims across Central Asia. In their world, these ideas are not so easily dismissed.

> One of the most monstrous plagues humanity has come up with is the idea of personal freedom…In many democratic societies the standard of living has been lowered below that of most farm animals…It was this freedom that led to the intense competition between capitalist states, leading to their colonization of the entire globe in order to steal natural resources and the material wealth of lesser nations, leading to the discrimination of other peoples contrary to all spiritual, moral and humanitarian values…Personal freedom and the principle of individualism transformed the societies of democratic states into such societies that they have fallen below the level of perverse and degrading animals.

While such screeching critiques against our own society are dismissed because we fail to see any similarity between rhetoric and reality, it must be said that this type of ruthless capitalism and cutthroat democracy exists and thrives in more places than we care to admit. The condition of society in Kyrgyzstan, Tajikistan and Uzbekistan acts as testimony to HT's accuracy: it is the common people of these states that are hopelessly lost in the shuffle, crushed between corrupt systems that offer no merit-based advancement and violent regimes that oppress with impunity. It is necessary to see past the lame Marxist rhetoric and accept the underlying social criticism. When done it becomes scarily clear that such areas of the world would willingly support radical movements that claim to be enemies of this degrading system and promise to strike at the true sources of the suffering. Ultimately, our main problem is we fail to see not only this inevitability, we fail to accept the emerging perception that labels the US as the ultimate true source: we are the fountainhead for global crypto-democracy. *Fostering Fundamentalism's* point simply places the blame for that perception on our failure to rectify our short-term hypocrisy in favor of our long-term security interests that can be established through *true* democratic engagement.

Even the old colonization rhetoric is not tired and empty in the Ferghana Valley. The HT skewers how easy it is to be magnanimous when a state such as the US has already rigged the global system in its favor. Gains dishonestly earned in the past have never come up for discussion within the international community. There has never been a discussion about undoing what were clear criminal acts. Time may soothe the conscience of the West but it inflames the disgust of radical Islamists. What this work has unfortunately shown is the resentful continuity that exists for fundamentalist Muslims: what we consider to be ancient history remains for them a present-day reality. US 'engagement' with the Ferghana Valley states has only proven to provide deepened repression as attacks against Islam have grown more violent.

All the while the US stands at the diplomatic podium and continuously praises 'small progress' and 'hopeful signs'. The US may very well be sincere in its desire to see repression in the Ferghana states end but its failure to be forthright in battling these perceptions, where we are perceived as doing nothing except pontificate, has been devastating to our overall image and has compromised our long-term security.

> Islam, which is the total antithesis of these democratic realities, is built upon Islamic doctrine, which demands the regulation of societal and governmental life to the commands and prohibitions of Allah. A person does not have the right to independently determine the societal system. He must live according to the system defined by Allah…
>
> Only Allah himself is legislator. Even if the entire Umma is finally united it is not within its rights to adopt a law if said law is against the original precepts first laid out by Allah himself. It would not matter even if every single individual Muslim came to an agreement about this law. None of these opinions carry any weight or value, they would not even carry the value of a mosquito's wings.

To western-educated people the above paragraphs possibly sound the most foreign. It is not so much that fundamentalist Islam can't accept the principle of freedom, as it supports freedom gained from people faithfully adhering to their religious beliefs. It is not so much that it can't accept economic activity, as it supports economic activity that advances the entire Umma and does not encourage personal gain. Fundamentalist Islam at its core is *anti-individualist*. In its worldview capitalism and western democracy are *only* about individualism. This is the chief conundrum we must overcome if we are truly to engage the Islamic world.

Why such hostility to personal freedom? As difficult as it is to accept in the West, such hostility is actually logically consistent with a core religious principle found in *both* Islam and Christianity: if man is born a sinner and doomed to sinful behavior when not spiritually constrained by the precepts of God, then his only salvation on earth is to adhere strictly to 'the commands and prohibitions' of the only true authority - God. Western individualism, more than anything else, appears to radical Islamists as the elevation of humanity to a position only God should occupy.

The Foundations of American Aggression[7]

> Islamic states must pay close attention to American pressure, which has the sole purpose of destroying Islam. This is a benefit which gives America weight and authority. America wishes to force the Muslim to believe in capitalism…
>
> Under the leadership of America all of the other capitalist nations strive for [close relations] with her. In order to achieve these relations they have begun to help America force her authority and her agents into the Islamic world…

The most influential weapon moving America's aggression forward is international sources of mass information. Rather than being a source of enlightenment these sources do nothing but confuse Muslims away from the true form of Islam while constantly accusing them of fundamentalism, extremism and terrorism. America's global information machine aims to portray Muslims as an enemy of all the world's peoples…All of this serves America's purposes as it functions to force Muslims to renounce their true faith and become slaves to capitalism…

Out of the many paths and methods used by America to further her aggression, one of the options is to put pressure on the pure hearts of the sons of Allah and force them to be silent, to make it so that no one remains who could speak the truth, to make them cower under waves of intimidation and terror…where blasphemy and the disgraceful behavior of non-believers become the norm.

These accusations accurately describe the political situation in the Ferghana states. They accurately portray life there for moderate and radical Muslims alike. Yet there are no direct entreaties mentioned against Karimov, Bakiev or Rakhmonov. It is America that is the focus of their ridicule, the central agent in a global conspiracy against their faith. It is possible to take this as empty braggadocio but this would be a mistake because it would be judging danger only along a spectrum of direct and provable culpability. Since HT is unlikely to be the direct initiator of terrorist action against the United States we have too simply discounted it as a serious threat. This work has proven how optimal the Ferghana Valley is as a transnational conduit and wayward haven for various radical groups and individuals. HT's rhetoric in this environment becomes more powerful because it reaches beyond the parameters and membership of its own organization. The West has unwisely chosen to ignore this reality. We are ignoring the power that HT's propaganda may have on the general perception amongst all Muslims, especially those exposed to radical groups and their teachings.

The essence of the American campaign is founded on four general principles: democracy, development, human rights and market economy. But before any of these principles are turned inside out and exposed for their falseness, it should be noted that the very essence of the American way is a refutation of logic…

The American doctrine is stuck between two antithetical thoughts: on the one hand is the idea of the importance of religion and that man should be subordinate to his faith, on the other hand is the simultaneous rejection of an almighty Creator and elevation of the human being to sole power and authority…

There either is a Creator, who created all people, all life and the world in which we live and to whom we are obligated to follow since after death we are held accountable by the manner in which we lived, or there is no Creator and therefore there should be no discussion whatsoever of a separation of religion from the state. If there is no Creator, religion should be ignored…

It is not acceptable to take a middle path and say that there is a Creator but that His existence does not hold great enough significance or meaning and should therefore occupy a smaller priority in the life and governance of man. Such thoughts simply do not sit in the minds of believing men and no one should be able to be comforted or satisfied with such a position.

Ironically enough, this type of logic should prove discomfortingly familiar to American Christian conservatives, a group not readily known for its empathy with radical Islam. Christian conservatives and Fundamentalist Muslims actually share many core ideas – the lessening of God in society, concern over the degradation of society's morals and the attempts by government to 'outlaw' the Creator just to name three. These ideas have formed the core of Christian conservative social criticism in America for over two decades. What should be most discomforting, however, to American Christians is the fact that of the two groups it is the radical Islamists who appear to be less hypocritical in their commitment to those beliefs.

What HT is actually criticizing is the saying, 'you can have your cake and eat it too'. This simple childish saying, which in more intellectual ways can be considered the motto of American ambition, comes off as base and contradictory when applied to religion. Radical Islamists are against 'compartmentalized religion', where you designate a certain time and place to be truly faithful and then at other times and in other places you can act more secular. This is a refutation of faith, pure and simple, to fundamentalist Muslims. Faith to the fundamentalist Muslim must always take first priority. Thus America comes across as not simply causing its own societal degradation but *forcing the degradation of other societies with whom it engages* because America is so adamant that its vision of secular-religious balance is superior and worthy of being mimicked. This seems especially apparent within the Ferghana Valley after 9/11, when all three states *increased* their oppression of Islam while aiding the US in its 'Global War against Terror' and sought and received simultaneous approval for phantom development along democratic lines. It is this seemingly brazen hypocrisy that makes HT look more sincere and legitimate in the valley not just in comparison with the local regimes but in comparison with the United States as well.

Freedom of Thought[8]

Freedom of thought in capitalist terms does not stop with the open criticism of government. Freedom of thought [in the West] also includes a permissiveness to immodesty, the rejection of the existence of God, drunkenness, debauchery and immoral sexual behavior, all of which are forbidden by God and only serves the purpose of annihilating Islamic values as it forces Muslims to act outside the boundaries of Islamic doctrine…

> Under conditions of Shari'a law a person does indeed have the right, even the obligation, to have his own personal opinion. There are instances where he is obligated to voice that opinion even if it runs counter to the views of the majority of Muslims. This is the case when a Muslim sees that someone speaks or does something against the principles of Allah. When this happens he must speak his mind with the purpose to end the apostasy that he has witnessed…
>
> But a Muslim is not allowed to speak something that is antithetical to Islamic doctrine. This includes all other modern ideologies, such as feminism, nationalism, secularism, capitalism, socialism, as they are not coincident with Islam…As the prophet himself said – *If your wishes will not let you stay in accordance with what I have brought (the Koran and Sunnah), then no one will be able to consider you a true believer.*

This is an entreaty to avoid hypocrisy and remain consistent in the belief that God comes before all else, even personal interest, desire and freedom. This type of criticism is possible against American society but it is even more cogent and consequential in the Ferghana Valley: there governments will openly display an absurdity of wealth and luxury while denying basic health and human services to the majority of the population. They do this while officially declaring allegiance to the ideas of freedom, liberty and prosperity and reaping the benefits personally of Western engagement. The Ferghana Valley states offer empty mimics of American ideology while America pretends to take them as progress. Radical Islamists notice how this game of diplomacy has succeeded in securing ever greater sums of aid from the United States while resulting in the violent contempt of Ferghana Valley governments toward their own peoples. To radical Islamists, however, the source powering this contempt is not Bishkek, Tashkent or Dushanbe. It is clearly Washington, DC.

Freedom of Acquisition[9]

> This freedom of capitalist states has expressed itself today in every sphere of life and has only brought enormous suffering to societies…it results in the expansion of criminal activity, the domination of the mafia, drug addiction, egoism and the ignoring of societal opinion…
>
> In addition this freedom has collected enormous wealth and riches and concentrated it into the hands of a few well-connected circles. These capitalists move to commandeer both domestic and foreign policy and ensure the priority of their own interests above society's…These capitalists do not pay attention to the suffering and pain of the people, to the bloodshed and criminality. They are only concerned with and obsessed by profit.

HT's portrayal of the freedom of acquisition perhaps sounds somewhat familiar to those who studied American economic history at the turn of the twentieth century, when barons such as Rockefeller, Morgan, Vanderbilt and Carnegie dominated American industry and political decision-making. But it does not accurately portray the American economic market today nor the state of American industry at present (Enron and Abramoff scandals notwithstanding). Therefore it again becomes too easy for many American policy circles and diplomatic agents to dismiss the relevance of HT's argument.

As mentioned before, however, it is not how accurate the criticism is in reality or how directly HT can connect it to America, it is *how the argument plays against perceived reality in the Ferghana Valley* that determines the long-term danger to the United States. The three preceding chapters have shown HT's critique to be an accurate reflection of state power and industry in Uzbekistan, Kyrgyzstan and Tajikistan. Thus while the evidence mounts higher and higher against the local regimes, HT is not losing the opportunity to show who these states are striving to resemble and to what philosophy they are adhering. It matters little that we do not see ourselves in such a light. HT does and is much more compelling a group to those who would decide to be violent against the 'enemies of Islam'.

The Politics of the Market Economy[10]

> In its open form the market economy brings only suffering to the Muslim. It contradicts Islam and therefore all Muslims are obligated to reject it…There is no doubt that these initiatives of America are in the first place taken against Muslims. Islam is the one ideology in the world today capable of offering an alternative to capitalism and democracy. It is the one true rival to these godless ideologies. It has been America's profound lack of true success that relegates it to name-calling and slander. Either America calls all those countries that are passionately committed to the doctrine of Islam terrorists, or it aims to subvert the Muslim community by making it subordinate to marionettes who rule with an iron hand over the people but in the end only answer and obediently listen to the American regime.

This paragraph accuses America of direct involvement in the oppression of the common Muslim citizen. As vicious and powerful the authoritarian autocrats of the Ferghana Valley may be, to radical Islamists they are nothing more than puppets on the strings of the puppeteer. America's Wonka Vision diplomacy made sure of that, as thousands were jailed and tortured while we praised the 'small progress' made by tyrants. To a victim the enabler is as responsible, in some ways more so, as the victimizer. In the volatile Ferghana Valley the HT helps too many victims see the United States as the one true enabler.

Terrorism[11]

> In 1979 the American and British intelligence services agreed that terrorism was the use of force against societal/civilian interests for the realization of political

> purposes...Out of all the definitions that emerged from that 1979 meeting it is clear that still today the defining process is not precise…
>
> In particular America has the habit of declaring every movement that acts in accordance with its wishes and priorities to be a movement of national resistance while all those that work against its priorities become terrorist organizations…America began to actively distribute its understanding of terrorism around the world and charging states as terrorist regimes when they were an obstacle to American policy. This was the case for North Korea, China, Iraq, Libya and Iran…
>
> On this foundation America is able to track and annihilate everyone and anyone that it characterizes as a terrorist…thus the very way that America defines terrorism allows it to be a tremendous strategic weapon in its arsenal to maintain dominance over the world…
>
> As such Islam was moved forward by America as a candidate to replace communism as its main enemy. Islamic countries simply occupy one of the most important regions of the world in which America wants to utilize its flexible interpretation of terrorism in order to push its influence and authority deeper…This is why today it is hardly a moment before a new Islamic movement appears in the Muslim world that America has not taken the opportunity to immediately label it a terrorist organization.

This criticism slices deep into the heart of America's foreign policy problem in the Ferghana Valley. If there was ever an opportunity for America to not show hypocrisy in principle it was in its dealings with Uzbekistan, Tajikistan and Kyrgyzstan. To this day none of the three are considered the 'front line' in the war against terror by the US government. As such it should have been less inclined to turn a blind eye to the repression and institutionalized devastation *that only succeeds in stoking the fires of radical Islam against America.*

As documented throughout the present work, radical Islamists do not take 'reform through engagement' as anything except a rationalization for oppression. Even worse, it is not so much that America comes to be regarded as hypocritical. It comes to be considered anti-Islam by design, with its ultimate objective to overrun and destroy the Muslim faith. This is the message HT succeeds in spreading across the valley despite its so-called commitment to 'non-violence'.

Fundamentalism[12]

> The characterization of certain Christian or Jewish groups as fundamentalist began in the West. Its point was to characterize these groups as being against science, industry and progress, all of which were supposedly delivered by the institutionalization of capitalism…

> The continuation of that characterization focuses largely on Islamic movements, made by western politicians and analysts, who also wish to equate Islamic fundamentalism with being a reactionary force, against any and all scientific and technical progress…As soon as such a designation is applied then it will lead to the exploitation of numerous cruel and ruthless measures taken against the offending groups…Every movement that acts against the aggressors who stole and continue to steal Muslim lands are called fundamentalists…Every Muslim who dies fighting enemy forces, who commits the selfless act of martyrdom, is a criminal suicidal psychopath…
>
> Muslims must understand that this use of the term fundamentalism is a political tool. It is not a true characterization of the movements of Islam, which are founded on faith in Allah…in judgment day and in divine predetermination…Islam is unique in that it is a holistic system of faith capable of deciding all the affairs of humanity…The naming of Islamic movements by America as fundamentalist is nothing but the struggle to prevent Islam from returning to prominence and is used by the West in purely strategic terms.

The only thing worth repeating here is how prominent America and the West figure in all of HT's formulations. Up to the present we have regarded groups operating in the Ferghana Valley as relatively weak and insular who may one day attempt regime overthrow at the local level but pose little international threat. But this is arguably naïve and short-sighted because there clearly seems to be very little within the official doctrine of HT focusing exclusively on the Ferghana states. The threat to Ferghana regimes may be slight not because of the weakness of radicals but because of their focus on the bigger prey they see as more directly responsible for their troubles: us.

The Aspiration of America to Annihilate Islam[13]

> America, at the present moment, wants to lead all nations, all laws, all religions and all people to capitalism. It wants to do this so that all people…rule themselves and their everyday lives according to the doctrine of capitalism…
>
> Only the Muslim community, the Umma, is truly able to stand against the world occupation of America's capitalism…America fears the rebirth of Islam on the world stage. Therefore it strives to prevent this reemergence at every turn. It strives to solidify its own authority and interests not only on its own territory but everywhere where human beings live.

In a local context so constrained by vicious oppression it is easy to see conspiracy theories where others see politics as usual. This was no different with HT, which looked on with knowing disdain in February 2004 when President Bush pushed forth a document at the G-8 summit called the 'Big Middle East Project'.[14] The official program was to encourage world consensus that pushed for political reform across the Middle East. It was titled 'Big' because it referred to the Middle East as a land mass stretching from Morocco all the way to Pakistan.

The project hailed in the West as a multilateral effort to promote global freedom was roundly hissed by radical Islamist groups who saw it as a brazen attempt at neo-colonialism.[15] It didn't help of course that not a single Arab state was consulted in the formation of the project. It was also not helpful that the text of the project included institutional reform of the Arab League so that it came into accordance with the 'global community's priorities'. To radical Islamists, this was simply an attempt by America to force its will and values ever further down the throat of the Muslim community.

While most Americans in one way or another sincerely believe in the superiority of democracy and capitalism, HT's astute analysis of the Big Middle East Project did expose some rather large and embarrassing loopholes which only served to deepen the animosity and suspicion of radical Islamists toward the West:

> It stands to remark that the desired political reform [pushed by America and forced on the G-8] does not include the right of local peoples to choose their own government, exercising their individual right of choice, if said choice will result in the coming to power of Islam or those who wish to see the rise of the Umma. This kind of change is unacceptable to the non-believer imperial powers, who strive to make sure their puppets remain in power. The purpose of political reform, therefore, is only to make these puppets come more into line with the points of view of the imperial powers. For them it is not important whether these puppets rule their peoples as tyrants or dictators. As long as they follow the puppeteer's doctrine, in general terms, the puppets are allowed to remain.[16].

As damning and abominable as that analysis is, the present work has shown time and again how often this criticism seems to ring true for people in the Ferghana Valley and how real-term consequences of American engagement there make it *easy* to believe.

Conclusion – The Bell of Muslim Unity Tolls for Thee

One of the major errors in analyzing radical Islamist groups in the Ferghana Valley has been the proclivity to focus on violent regime overthrow and the likelihood of particular groups' actual involvement in violence. While these priorities are by no means irrelevant or not worthy of study, this chapter has asked that the realm of investigation be widened so that equally dangerous but more subtle threats become visible. Subtlety in this case does not signify weakness – in the post-9/11 world the West has learned that sometimes the threats that are the least visible and take the longest to come to fruition can often be the most deadly.

This is especially so in the case of the Hizb ut-Tahrir. While long on the radar screen their so-called commitment to non-violence has been taken as more important than their actual beliefs or the manner in which they express them. The

HT is not at present able to mount an armed force against any of the three Ferghana states. It is also not ready to physically pursue its own international agenda resulting in various terrorist acts, if it were so inclined. But what is obvious is that HT proselytizes a vision of the world that fosters the clash of civilizations hypothesis. It divides the world into believers and non-believers and pushes the idea that non-believers are waging a holistic war against Islam that includes not just military operations but also economic, political, cultural, religious and sexual warfare. In this war where the West is bent on destroying Islam there is one obvious main antagonist. It is not Kyrgyzstan, nor is it Tajikistan or Uzbekistan. Without doubt or ambiguity the main devil for the HT is America.

What this chapter does is respectfully admonish analysts for being too literal in their approaches on conflict and too short-sighted in their understanding of threats. The Ferghana Valley is a perfect cauldron to mix various disaffected, repressed and isolated factions into a volatile cocktail waiting for the right fuse. The HT may not be the very fuse to light the bomb. But it is certainly providing the atmosphere in which the wick can be made, one radical Islamist string at a time. If there is any doubt of this then I simply remind the reader of the Hizb ut-Tahir's own call to Muslims. The West should pay attention to this call because it is becoming more and more certain that Muslims in the Ferghana Valley certainly are:

> Today you stand face to face before your ultimate obligation – the defense of your religion and your knowledge. You are the guard over your entire community…
>
> Die for truth and justice! The time has come not to differentiate between life and death. America and the godless West, your rulers, including the consultants and advisors and businessmen and politicians, they are all in a bottomless pit…Those who fight for Islam will be a son of Islam and you will find yourself on the one true path!
>
> It is demanded of every Muslim who believes in Allah and the Prophet, who believes in the religion given to Muhammad, to stand and find oneself on the true path. There is no alternative to this question as there is no possibility for any Muslim to remain neutral on these issues.[17]

Just as democratic development and economic prosperity was conveniently considered a relative thing by the United States when engaging the Ferghana Valley states, it is apparently equally relative how the HT defines non-violence. When analyzing the content and underlying meaning of HT's ideology it is indisputable that its 'non-violence' is quite deadly for Americans. It is high-time we recognize.

---

[1] There is a wealth of material written about the HT by the HT. Unfortunately, most of it is not translated into English. I have taken that duty upon myself in this chapter. All quotes and excerpts taken from the Hizb ut-Tahrir are translated personally by me, from the Russian, Arabic and at times Uzbek. Any mistakes or misinterpretations, therefore, will have to be my responsibility.

[2] Ariel Cohen, 'Radical Islam and US Interests in Central Asia', Testimony before the Subcommittee on Middle East and Central Asia, Committee on International Relations, US House of Representatives, 29 October 2003.

[3] International Crisis Group, 'The IMU and the Hizb ut-Tahrir: Implications of the Afghanistan Campaign', Central Asia Briefing, Osh/Brussels, 30 January 2002.

[4] Ariel Cohen, 'Radical Islam and US Interests in Central Asia', Testimony before the Subcommittee on Middle East and Central Asia, Committee on International Relations, US House of Representatives, 29 October 2003.

[5] Ariel Cohen, 'Radical Islam and US Interests in Central Asia', Testimony before the Subcommittee on Middle East and Central Asia, Committee on International Relations, US House of Representatives, 29 October 2003.

[6] Hizb ut-Tahrir, 'Demokratiya Sistema Neveriya', http://www.hizb-ut-tahrir.org/russian/htm/democracy.htm, accessed 12 December 2004.

[7] Hizb ut-Tahrir, 'Opori Amerikanskoi Aggressii', http://www.hizb-ut-tahrir.org/russian/htm/esdarat.htm, accessed 23 June 2005.

[8] Hizb ut-Tahrir, 'Opori Amerikanskoi Aggressii', http://www.hizb-ut-tahrir.org/russian/htm/esdarat.htm, accessed 30 August 2005.

[9] Hizb ut-Tahrir, 'Opori Amerikanskoi Aggressii', http://www.hizb-ut-tahrir.org/russian/htm/esdarat.htm, accessed 5 March 2005.

[10] Hizb ut-Tahrir, 'Opori Amerikanskoi Aggressii', http://www.hizb-ut-tahrir.org/russian/htm/esdarat.htm, accessed 23 June 2005.

[11] Hizb ut-Tahrir, 'Demokratiya Sistema Neveriya', http://www.hizb-ut-tahrir.org/russian/htm/democracy.htm, accessed 12 December 2004.

[12] Hizb ut-Tahrir, 'Demokratiya Sistema Neveriya', http://www.hizb-ut-tahrir.org/russian/htm/democracy.htm, accessed 16 February 2005.

[13] Hizb ut-Tahrir, 'Demokratiya Sistema Neveriya', http://www.hizb-ut-tahrir.org/russian/htm/democracy.htm, accessed 8 October 2005.

[14] Hizb ut-Tahrir, 'Otveti na Voprosi', http://www.hizb-ut-tahrir.org/russian/htm/democracy.htm, accessed 4 February 2005.

[15] Hizb ut-Tahrir, 'Otveti na Voprosi', http://www.hizb-ut-tahrir.org/russian/htm/democracy.htm, accessed 4 February 2005.

[16] Hizb ut-Tahrir, 'Otveti na Voprosi', http://www.hizb-ut-tahrir.org/russian/htm/democracy.htm, accessed 4 February 2005.

[17] Hizb ut-Tahrir, 'Otveti na Voprosi', http://www.hizb-ut-tahrir.org/russian/htm/democracy.htm, accessed 4 February 2005.

Chapter 8

Conclusion: Building New bin Ladens

Democracy is hypocrisy without limitation.
Iskander Mirza (1899-1969, President of Pakistan)

The end of the Cold War was for many in the West a confirmation of the universality and superiority of both democracy and capitalism. The control of individualism and the management of human ambition were not only lesser ideologies, they were actually futile as peoples all over the globe stepped forward wanting to become part of the cradle of democracy. From the very beginning of the post-Cold War era, therefore, the United States proclaimed to operate on two general principles:

1 – Achieving balance between initiatives that address immediate threats and programs that promote lasting generational change.
2 – Pursue selective engagement with those states that are truly willing to reform.[1]

These two principles were never consistently and coherently applied even before 9/11, but 9/11 itself proved to be a tipping point: it marked a shift elevating short-term strategic interests high above long-term 'generational' reform. There was little opportunity to criticize this shift immediately following the Twin Towers disaster as it rightfully seemed awkward and indecent to complain about security methods when the United States had been so brazenly attacked.

Please remember that the purpose behind this work is not partisanship. It is not to vent diatribes against the right, hawks, the military-industrial complex or any group that believes security should first and foremost always be the top priority for a state. The flaws discussed within are endemic to both the Democratic and Republican parties and in some ways reflect a flaw in the philosophical reasoning with which we approach international relations and develop state foreign policy. This work exists to call attention to the fact that our stalwart pursuit of present-day terrorism ('the programs that address our immediate threats') has been done in a way that will not only *NOT* produce the 'lasting generational change' that we say is ultimately most important but will in fact enable the reemergence of transnational terrorist groups and give them the arguments needed to attract new legions to the cause of bringing down the dominance of the West.

Our hypocrisy *of our own professed foreign policy* creates new generations of terrorists whose one true goal will not be in attacking the local regimes with which we 'selectively engaged' but will be in looking to cause pain and suffering against Americans. They will seek out this retribution because the United States was not simply indifferent to their own repression. It will appear to them by all objective methods of investigation to be something we tacitly endorsed and consistently supported. The Ferghana Valley is just the first and perhaps the best positioned of many other potential sites around the globe.

Why is this argued? I argue this because we have blindly allowed our belief in immediate security to be embodied by a system dominated by politicians and diplomats who have been willing since 9/11 to overlook democratic regression in places like the Ferghana Valley. This democratic regression is considered a 'necessary evil' to achieve progress in the present Global War against Terror. This is not an argument, however, simply railing against immediate security concerns and lamenting over idealized visions of principles like freedom and liberty: *Fostering Fundamentalism* is bringing to light the manner in which the United States provided aid to the region and how such manners compromise our security in the more important long-term. While the local regimes increased their repression we looked duplicitous and self-serving.

To conduct the war against terror up to now the US government has relied mostly on its foreign military assistance programs, which are defined as a method for 'friends and allies to acquire US military equipment, services, and training for…legitimate self-defense and for participation in multinational security efforts'.[2] In trying to use these programs more effectively to forge its international coalition against terror, the US has modified its military assistance stance in three ways:

- Changing the legal regime to push arms transfers to foreign nations.
- Granting military assistance to several states directly involved in the war against terror.
- Increasing and expediting counterterrorism assistance and general military aid to other countries around the world.[3]

Unfortunately, how this has played out in real-terms is an uncomfortable softening and arguably even a violation of formal US law. Section 502B of the Foreign Assistance Act of 1961 forbids the transfer of assistance to governments that engage in a 'consistent pattern of gross violations' of human rights.[4] Under 'extraordinary circumstances' the president has the right to waive these restrictions, which President Bush has done rather liberally since 9/11. The State Department also began lifting sanctions as part of its post-9/11 foreign policy, while the Defense Security Cooperation Agency, which handles foreign military sales within the Department of Defense, established a 'war room' so as to speed up sales approvals and more quickly expedite transfers.[5] All of this pushes radical Islamist circles to one conclusion – we want to help kill Muslims faster. Whether this is actually an accurate reflection of reality matters little. In such isolated places suffering from severe repression and isolation like the Ferghana Valley, perception is more important.

Our press conferences and diplomatic communiqués that offer up 'hope for the future' or 'contentment with recent progress' all ring hollow and deadly to local groups *having to live amongst the repression*. Even when we voice concern, as Assistant Secretary of State Lorne Craner did in 2004, it comes off muted and lacking authority:

> We know that while there is no justification for terrorism, repressive societies without economic development and that [practice] social exclusion have been breeding ground for terrorists. That is a simple fact. We don't want to see that continue.[6]

This is well said but how does that play on the ground as government forces in Bishkek, Tashkent and Dushanbe use weapons given by the US military at an 'expedited pace' to shoot and kill Muslims? We must stop undervaluing the impact of perception on our long-term security. What becomes of all the moderate Muslims in the Ferghana Valley imprisoned and tortured? What becomes of their family and friends who *lived that reality* while listening to the US talk of being pleased overall by the 'careful progress' achieved in the region through American engagement? While this is not necessarily revelatory and the US is certainly not the first country to engage foreign partners on security matters while having reservations about the nature of the regime they are involved with, the problem is that these regimes helping us fight terrorism today are actually causing the terror problems we will have to fight tomorrow.[7] Even the NGO community within the Ferghana Valley itself, explained best by the director of the Coalition for Democracy and Civil Society in Kyrgyzstan, sees the situation in a scarily compromised and pragmatic way:

> I regard the presence of the US in [and assistance to] our region…as very positive as *there will be no end to terrorism.* Therefore it's hard to say the US will leave the region in the next 10 years.[8]

Since repression within the Ferghana states is so horrendous, and will continue to be despite our efforts and assistance, we will be forced to remain in the area. In short, our assistance to fight terrorism in the valley only feeds more terrorism and thus grass-roots organizations at the local level have come to approve of this development because it means we won't abandon them. The environment has become so perverse that even local civil society NGOs are finding ways to positively view the strategic value of terrorism.

Even more disturbing, the hundreds of millions of dollars poured into the region since 9/11 have produced little significant results while untold amounts have conveniently ended up in the hands of elites. Yet we persistently proclaim success and progress. This is what the present work calls Wonka Vision diplomacy and it is pervasive throughout the entire US government when justifying its fight against terror. Elizabeth Jones, Craner's boss for all intents and purposes, showed how deep Wonka Vision runs when testifying in 2003 before the Subcommittee on the Middle East and Central Asia on Capitol Hill:

Rep. William Janklow (SD): In reality, where are we really making headway in terms of the people themselves thinking that there is a friend out there that is us? Don't they really associate us with the actions of their governments? Aren't they the same as people elsewhere around the world?

Jones: Actually, on the contrary…Were it not for the support that the United States provides these organizations, they would not be able to function. And as tough as it is, they are the ones who come to us and say, please don't abandon us; you are the ones that are helping us to keep going…

What we have done in the Ferghana Valley is picked a few towns or villages where we have put in very comprehensive assistance programs. So that we don't have just some agricultural reform programs through the Ferghana Valley, and some education programs and some exchange programs and some judicial reform programs, we have put a whole comprehensive package in place in several towns and villages where we thought the danger of Islamic extremism was the greatest…And it is job generation, education and a lot of discussion with civil society and with government leaders about the importance of not using repressive police tactics as a way to control Islamic extremism.[9]

Jones understanding of her own Bureau's impact is inflated at best and wholly disjointed with reality at worst. There is indeed a sense of abandonment in the valley. Islamic extremism has not been countered by the Bureau's comprehensive programs in a few towns and villages because the valley itself is regularly inundated by outside groups and propaganda, further solidifying the valley's conduit reputation. Government leaders have not heeded pleas to lessen police violence and have received no reprimand, no negative policy consequences as a result of their defiance. Representative Janklow's concern was dead on-target: they do see us as a part of their own repressive government. In the Ferghana Valley, the US is no friend.

The repressive methods employed by Karimov, Rakhmonov and Askaev/Bakiev have also not made the Ferghana Valley less volatile or more secure. The Ferghana Valley has become a perfect nexus for harboring radicals, either in groups or as individuals: the borders between the three states are still porous and poorly guarded. The valley itself remains a largely ungoverned region where secret unfettered movement from city to city remains easy. The classic terrorist cell system, small and isolated from each other so as to prevent intelligence penetration, has only deepened in the valley since 9/11, not weakened. The valley has long been a transnational conduit for narcotics and slave smuggling. Transforming these conduits so that they can also effectively allow for the flow of radical Islamists is not difficult. If Osama bin Laden wanted to remain forever hidden, he should have fled north from Afghanistan to the valley, where the US does not have as much access while the local regimes use blunt but inefficient methods to antagonize Muslims. Hiding in the valley away from international eyes is in fact easier *because* of the repressive hold the local regimes pretend to maintain over their sovereign territory. It may seem like a paradox but the fact of the matter is that they may repress the valley but they do not hold it.

At the moment US diplomacy seems to live in a bubble impervious to understanding that Uzbek, Tajik and Kyrgyz repression against Islam does *NOT* just damn those regimes. Since 9/11 a dramatic shift has occurred in public perception within the valley that sees the United States as a partner in that repression. After all, it hasn't gotten better, American diplomatic rhetoric notwithstanding, the repression has gotten worse. As Assistant Secretary Craner said on a different occasion:

> The region I have been to most in this job is Central Asia. I have visited no country more often than Uzbekistan. I am bringing a message to the governments and I hope to the people of how much we care about advancing human rights and democracy in Central Asia. And I am also bringing a lot of help, material help, to actually be able to do that. People in the region should understand that.[10]

Poor oppressed peoples never see coincidence. They see conspiracy. The American government would do well to remember that next time it is in front of the podium and allow that knowledge to be reflected in its actual policy deeds. Talking the talk without walking the walk has been disastrous in the ability to win the war against terror. Why? This is because slavish commitment to the utility of short-term strategy compromises long-term success *over the same strategic interest*. While I do not doubt the sincerity of polished diplomats like Craner, Jones, Karen Hughes, Condoleeza Rice and Donald Rumsfeld, there are consequences to American security when they cannot see the fatal contradictions which all Muslims in the Ferghana Valley hear when our diplomacy speaks of progress and engagement.

In addition to this diplomatic aspect, the few warnings that have emerged from the scholarly community about radical Islamists in the valley continue to be overshadowed by counter-analyses that emphasize the relatively small size or lack of military capability among radical groups to attack local regimes. Appearing before the same subcommittee in 2003 as Jones were four leading scholars on the region. Embedded deep within their official testimony were comments that the present work finds so disconcerting.

> DR. ARIEL COHEN: While reports of increasing Hizb activity abound, the extent to which local Hizb activities are part of a coordinated global plan is still unknown, just as the question of whether every region and country has an autonomous leadership that defines programs and sets deadlines remains unanswered. Hizb is rumored to be operating on a thirteen-year grand plan, but if it exists, this program is still unknown…
>
> In fighting the War of Ideas, the Bush administration must reevaluate, revive, and upgrade its public diplomacy tool box, as well as invent new specific tools

for fighting aggressive, anti-Western sentiment among fundamentalist groups and regimes, which support and tolerate them.

DR. STEPHEN BLANK: Virtually every writer on Central Asia has postulated that the combination of ubiquitous misrule, corruption, poverty and repression there runs the risk of encouraging opposition groups to gravitate toward Islamic parties and movements for want of any other option. The lack of an option is therefore allegedly due to the fact that the regimes there have stifled all other opposition movements. Hence Islamic movements, which are generally and inherently underground operations, are lest as the only force capable of arousing opposition to this misrule. Alternatively this repression and misrule stimulates this gravitation to Islamic parties because only they have the most coherent and resonant message that the population can assimilate in terms it understands.

This conclusion emerges because it is assumed that all other avenues of political expression are closed off due to repression, socio-economic decline, environmental degradation, the breakdown of social norms through crime, corruption, and drugs, ethnic cleavages, and/or the absence of a genuine civil society. Hence Islamic parties and movements that supposedly speak to the populace in their own language are left by default as the only alternative.

DR. MARTHA BRILL OLCOTT: One of the most frustrating aspects of working in this region is how limited our levers are to influence developments in the region. Given the levels of US spending in the area, our greatest area of influence is in the area of security, and when we tackle state building issues, especially issues of political reform, it is much harder for our voice to be heard, given how limited the funds are that we are spending in this area in any of these countries.

DR. FIONA HILL: I believe that the US should encourage more programs to expand political participation…We need to bring extremist groups out of the shadows by encouraging religious education at all levels, allowing people to make their own decisions and to publicly debate social problems in mainstream settings so that these issues will be taken out of the domain of radical interpretation.[11]

Clearly the scholarly community can see the potential dangers but has not, for whatever reason, causally tied the exacerbation of those dangers directly to the contradictions and hypocrisy of American foreign policy. They are still focused largely on the repression initiated by local governments and at least somewhat ignoring the fact that such repression is violently enabled by our own 'selective engagement'. This work is hopefully the first in what will be a long line of serious investigations into the long-term security impact of 'diplomacy for show' and the acceptance of cosmetic results when substantive progress is needed.

It is imperfect diplomacy when the US allows democratic principles to be undermined in areas already sensitive to radical Islamic thought and then goes on to proclaim success in establishing a foundation for democratic principles. This only helps fuel Muslim disgust toward American ideals and values. That imperfect diplomacy mutates into Wonka Vision democracy when those compromises are not only openly justified but are falsely trumpeted as the causes of progress. This mutation helps undermine the Global War on Terror not just today but *generationally*. And while everyone knows that a complete end to terrorism is impossible and that terrorists will always exist, this is not a rationale for tolerating policy that in and of itself creates future terrorists and provides them with damning evidence as to why they should act against the United States.

Conditions of poverty, repression, censorship, corruption and injustice truly do lead to radicalization. Unfortunately, our engagement with the Ferghana states has not lessened these qualities. In many ways our engagement has brought about their intensification. Yet we have not acted against this heinous coincidence and have spoken out against it only weakly and lamely. We do this because we stubbornly adhere to the misplaced thinking that a short-term betrayal of true democracy abroad is sometimes necessary for making sure terror does not strike at home. This thinking is not necessary or acceptable because it is wrong.

Duplicity and selfishness have long been common characteristics in foreign-policy propagation; one only need read Machiavelli's *The Prince* to know just how long such subterfuge has been implicitly if not explicitly endorsed by the international community. The problem elaborated here is that our duplicity and selfishness *do not accomplish our security objectives*. As Machiavelli himself wrote, if it were possible to be a beacon of hope and justice and have all of the people love you while maintaining power and security, then it would be best to be such a beacon. The problem for Machiavelli was reality didn't seem to provide many such opportunities to be positive *and* powerful. What the US needs to realize is that long-term victory in the Global War against Terror is the very opportunity that Machiavelli deemed unlikely: an opportunity where the steadfast commitment to justice and liberty, without any space for hypocrisy or contradiction, actually succeeds in achieving greater success and gaining greater security. The short-term temptations intrinsic to this war should be ignored for the long-term strategies that will ultimately produce the greatest results. We are indeed served best by being both positive and powerful.

Our betrayal of democracy is not just a betrayal of the people of the Ferghana Valley and all those in places just like it. Our betrayal of democracy is a betrayal of the very essence of our hegemony. It destroys the one filament that could arguably light and justify American dominance: unilateralism would be acceptable if it was selflessly committed for the true betterment of others. Hegemony would be allowable if it was used to sincerely expand the principles of freedom, liberty, prosperity, dignity and justice. This type of unilateral hegemonic power would be capable of gaining world acquiescence because it would not be about American interests but human ones. This is what can bring victory over terror: a belief by the world community that this truly is our motivation and our inspiration. That belief only comes when the world community consistently sees us walk the talk we so eloquently proclaim about liberal democracy.

But that is not our hegemony today nor is it the end result of our unilateralism. Convenient strategic pragmatism infuses our foreign policy and allows the government to justify and rationalize its excesses and failures. This is a refutation of the true American ideal. It is a rejection of the purpose American ideology was supposed to serve in the global arena. It is not too late to correct this policy flaw. Before we can correct it, however, one thing is certain: both diplomatically and scholarly we must stop ignoring the signs that we are blindly building new bin Ladens.

---

1 Bureau of European and Eurasian Affairs, 'US Government Assistance to and Cooperative Activities with Eurasia', Fiscal Year 2002, Washington DC.

2 Human Rights Watch, 'Dangerous Dealings: Changes to US Military Assistance After September 11', Vol. 14, No., 1(G), February 2002.

3 Human Rights Watch, 'Dangerous Dealings: Changes to US Military Assistance After September 11', Vol. 14, No., 1(G), February 2002.

4 Human Rights Watch, 'Dangerous Dealings: Changes to US Military Assistance After September 11', Vol. 14, No., 1(G), February 2002.

5 Human Rights Watch, 'Dangerous Dealings: Changes to US Military Assistance After September 11', Vol. 14, No., 1(G), February 2002.

6 International War and Peace Reporting, 'Will US Policy Backfire in Central Asia?' 30 March 2004.

7 International War and Peace Reporting, 'Will US Policy Backfire in Central Asia?' 30 March 2004.

8 International War and Peace Reporting, 'Will US Policy Backfire in Central Asia?' 30 March 2004.

9 Hearing before the Subcommittee on the Middle East and Central Asia of the Committee on International Relations, House of Representatives, 'Central Asia: Terrorism, Religious Extremism, and Regional Stability', 108th Congress, 1st Session, Washington DC, 29 October 2003.

10 International War and Peace Reporting, 'Will US Policy Backfire in Central Asia?' 30 March 2004.

11 Hearing before the Subcommittee on the Middle East and Central Asia of the Committee on International Relations, House of Representatives, 'Central Asia: Terrorism, Religious Extremism, and Regional Stability', 108th Congress, 1st Session, Washington DC, 29 October 2003.

Appendix

Relevant Web Links for Central Asia

Academic Institutions and Educational Foundations

The Caucasus and Central Asia Program at University of California – Berkeley:
http://ist-socrates.berkeley.edu/~bsp/caucasus/

Central Asia and the Caucasus at Columbia University:
http://www.columbia.edu/cu/sipa/REGIONAL/ECE/CACR/

Central Asia and the Caucasus – Institute for Central Asian and Caucasian Studies, Sweden/Institute of the Strategic Studies of the Caucasus, Azerbaijan:
http://www.ca-c.org/

Central Asia-Caucasus Institute – Johns Hopkins University:
http://www.cacianalyst.org/

Central Eurasian Studies Society – Harvard University:
http://cess.fas.harvard.edu/index.html

Davis Center for Russian and Eurasian Studies – Harvard University:
http://daviscenter.fas.harvard.edu/

The Eurasia Foundation:
http://www.eurasia.org/

The Inner Asian and Uralic National Resource Center at Indiana University:
http://www.indiana.edu/~iaunrc/caweb.html

Institute for Defense Analyses:
http://www.ida.org/

Institute for Foreign Policy Analysis:
http://www.ifpa.org/home.htm

International Research and Exchanges Board (IREX):
http://www.irex.org/

Military Education Research Library Network:
http://merln.ndu.edu/

National Defense University:
http://www.ndu.edu/

The Open Society Institute and Soros Foundations Network:
http://www.soros.org/

United States Institute for Peace:
http://www.usip.org/

Woodrow Wilson International Center for Scholars:
http://wwics.si.edu

Policy Think Tanks

American Enterprise Institute:
http://www.aei.org/

The Carnegie Endowment for International Peace:
http://www.carnegieendowment.org/

The Center for Defense Information:
http://www.cdi.org/

Center for Strategic and International Studies:
http://www.csis.org/

Council on Foreign Relations:
http://www.cfr.org/

The Eisenhower Institute:
http://www.eisenhowerinstitute.org/

Foreign Policy in Focus: A Think Tank Without Walls:
http://www.fpif.org/

Global Policy Forum:
http://www.globalpolicy.org/

Global Witness: Resources, Conflict and Corruption:
http://www.globalwitness.org/

The Heritage Foundation:
http://www.heritage.org/

The Hudson Institute:
http://www.hudson.org/

Kreml.org – The Political Expert Network:
http://www.kreml.org/

The National Endowment for Democracy:
http://www.ned.org/

Rand Corporation Center for Russia and Eurasia:
http://www.rand.org/nsrd/cre/

American and Intl. Governmental + Non-Governmental Organizations

Amnesty International:
http://www.amnesty.org/

The Asian Development Bank – Central Asian Regional Economic Cooperation:
http://www.adb.org/CAREC/default.asp

Coalition for Democracy and Civil Society:
http://www.ngo.kg/

Commission on Security and Cooperation in Europe:
http://www.csce.gov/

Eurasianet:
http://www.eurasianet.org/

European Union External Relations with Eastern Europe and Central Asia:
http://europa.eu.int/comm/external_relations/ceeca/

Human Rights Watch:
http://www.hrw.org/

International Crisis Group:
http://www.crisisgroup.org/

IFEX – International Freedom of Expression eXchange:
http://www.ifex.org/

UNESCO in Central Asia:
http://www.unesco.org/webworld/centralasia/

United Nations Integrated Regional Information Networks:
http://www.irinnews.org/homepage.asp

US Agency for International Development:
http://www.usaid.gov/

US Department of State:
http://www.state.gov/

Local Central Asian and Islamic Sources

Ar-Namys Party:
http://www.ar-namys.org/

Birlik Party:
http://www.birlik.net/

Central Asian Gateway:
http://www.cagateway.org/

The Central Asia Non-governmental Organization Network:
http://www.cango.net.kg/

Erk Democratic Party of Uzbekistan:
http://www.euronet.nl/users/turkfed/erk.html

Ferghana Online:
http://www.ferghana.uz/ru_home.html

The Government of the Republic of Kyrgyzstan:
http://www.gov.kg/

The Government of the Republic of Uzbekistan:
http://www.gov.uz/

Hizb ut-Tahrir:
http://www.hizb-ut-tahrir.org/

Jalal-Abad Oblast:
http://www.djalal-abad.gov.kg/

Muslim Uzbekistan:
http://archive.muslimuzbekistan.com/

Osh Oblast:
http://www.osh.gov.kg/

Presidential Website of the Kyrgyz Republic:
http://www.president.kg/president/

Press Service of the President of the Republic of Uzbekistan:
http://www.press-service.uz/

The Republic of Tajikistan:
http://tajikistan.tajnet.com/

Sunshine Uzbekistan:
http://www.sunshineuzbekistan.org/

Umid World – Portal to Uzbekistan:
http://www.umid.uz/Main/

The World Bank:
http://www.worldbank.org/

Journalism and Media Sources

Asia Times Online:
http://www.atimes.com/

Central Asia Newsnet:
http://www.centralasianews.net/

Ferghana Central Asian Information Agency:
http://www.ferghana.ru/

Institute for War and Peace Reporting:
http://www.iwpr.net/

Journal of Turkish Weekly:
http://www.turkishweekly.net/

Kyrgyzstan Newsnet:
http://kyrgyzstannews.net/

Kyrgyz Press:
http://www.kyrgpress.org.kg/

Media Association of the Ferghana Valley:
http://eng.fergana.org/

News Central Asia – The Voice of Greater Central Asia:
http://www.newscentralasia.com/

Radio Free Europe/Radio Liberty:
http://www.rferl.org/

Tajikistan Newsnet:
http://tajikistannews.net/

The Times of Central Asia:
http://www.timesca.com/

Voice of America:
http://www.voanews.com/

Bibliography

Accord: An International Review of Peace Initiatives, 'A Chronology of Events in Tajikistan', http://www.c-r.org/accord/Tajik/accord10/chronology.shtml, (Washington DC: 14 December 2004).

Adams, Laura, 'Cultural Elites in Uzbekistan: Ideological Production and the State', In *The Transformation of Central Asia: States and Societies from Soviet Rule to Independence*, edited by Pauline Jones Luong, 93–119, (Ithaca: Cornell University Press, 2003).

Akbarzadeh, Central Asia-Caucasus Analyst, 'Is Islam a Threat to Stability in Central Asia?' http://www.cacianalyst.org/view_article.php?articleid=188, (Johns Hopkins University: 12 July 2004).

Akbarzadeh, Shahram, *Uzbekistan and the United States: Authoritarianism, Islamism and Washington's New Security Agenda*, (Zed Books, 2005).

——'A Note on Shifting Identities in the Ferghana Valley', *Central Asian Survey* 16, no. 1: 65–68 (1997).

Aljazeera New Service, 'Bereaved Uzbek Mother Gets Prison'. http://english.aljazeera.net/NR/exeres/9B813F38-57CC-4DF0-8147-2D3F38B04316.htm, (Qatar: 21 December 2004).

Allworth, Edward, ed., *Central Asia: 130 years of Russian dominance*, (Durham: Duke University Press, 1994).

Anderson, John, 'Constitutional Development in Central Asia', *Central Asian Survey* 16, no. 3: 301–320 (1997).

Arman, Kambiz and Nazar Nazarov, Eurasianet, 'Tajik President Seems to Resist Pressure for Political Liberalization'. http://www.eurasianet.org/departments/civilsociety/articles/eav041905.shtml, (Washington DC: 23 August 2005).

Atkin, Muriel, 'Thwarted Democratization in Tajikistan', In *Conflict, Cleavage, and Change in Central Asia and the Caucasus*, edited by Karen Dawisha and Bruce Parrott, 277–211, (Cambridge: CambridgeUniversity Press, 1997).

Bacevich, Andrew and AJ Bacevich, *The New American Militarism*, (Oxford: Oxford University Press, 2005).

Benjamin, Daniel and Steven Simon, *The Next Attack: The Failure of the War on Terror and a Strategy for Getting it Right*, (Henry Holt & Co., Inc, 2005).

Beshimov, Erdin, Eurasianet, 'Akayev: We do Democracy the Kyrgyz Way', http://www.eurasianet.org/departments/insight/articles/eav101504.shtml, (Washington DC: 23 March 2005).

The Bureau of Democracy, Human Rights, and Labor, 'Mission Statement', http://www.state.gov/g/drl/democ/, (Washington DC: 15 March 2005).

——'US Human Rights and Democracy Strategy', http://www.state.gov/g/drl/democ/, (Washington DC: 15 March 2005).

——'Fact Sheets', Department of State, United States of America, (Washington DC: 1 June 2004).

Bureau of European and Eurasian Affairs, 'US Government Assistance to and Cooperative Activities with Eurasia', http://www.state.gov/p/eur/rls/rpt/37670.htm, (Washington DC: 13 December 2005).

——'US Policy and the Organization for Security Cooperation in Europe'. http://www.state.gov/p/eur/rls/rpt/2004/31935.htm, (Washington DC: 31 October 2005).

——'US Government Assistance Report 2004'. http://www.state.gov/p/eur/rls/rpt/37677.htm, (Washington DC: 11 February 2005).

——'US Government Assistance to and Cooperative Activities with Eurasia'. http://www.state.gov/p/eur/rls/rpt/37645.htm, (Washington DC: 27 August 2004).

——'Frequently Asked Questions about US Policy in Central Asia'. http://www.state.gov/p/eur/rls/fs/15562.htm, (Washington DC: 28 July 2004).

——'US Government Assistance to and Cooperative Activities with Eurasia'. http://www.state.gov/p/eur/rls/rpt/37670.htm, (Washington DC: 24 June 2004).

Byman, Daniel and Matthew Waxman, *Dynamics of Coercion: American Foreign Policy and the Limits of Military Might*, (Cambridge: Cambridge University Press, 2002),

Callahan, Patrick, *America's Foreign Policy: Theories of America's World Role*, (New York: Longman Publishing, 2003).

Carothers, Thomas, *Aiding Democracy Abroad: The Learning Curve*, (Washington DC: Carnegie Endowment for International Peace, 1999).

——*Critical Mission: Essays on Promoting Democracy*, (Washington DC: Carnegie Endowment for International Peace, 2004).

Central Asia-Caucasus Analyst, SAIS, 'OSCE Pleased with Kyrgyz Presidential Ballot', http://www.cacianalyst.org/view_article.php?articleid=3521, (Johns Hopkins University: 15 November 2005).

——'Assistant Secretary of State Jones on US Policy in Central Asia'. http://www.useu.be/Terrorism/USResponse/Dec1401JonesUSPolicyCentralAsia.html, (John Hopkins University: 12 April 2004).

——'Kyrgyz Authorities Arrest Ex-speaker of Parliament'. http://www.cacianalyst.org/view_article.php?articleid=3433, (Johns Hopkins University: 19 August 2005).

——'Kyrgyz Deputy Suggests Year-long Moratorium on Protests'. http://www.cacianalyst.org/view_article.php?articleid=3753, (Johns Hopkins University: 9 November 2005).

——'Kyrgyz Forces Break up the Protest'. http://www.cacianalyst.org/view_article.php?articleid=3429, (Johns Hopkins University: 18 August 2005).

——'Lawmaker Slams Marred Balloting in Kyrgyzstan, Tajikistan'. http://www.cacianalyst.org/view_article.php?articleid=3138, (Johns Hopkins University: 23 September 2005).

——'Media and Elections in Central Asia'. http://www.cacianalyst.org/view_article.php?articleid=2707, (Johns Hopkins University: 24 September 2004).

——'Rakhmonov's Election Maneuvers and Tajikistan's Future'. http://www.cacianalyst.org/view_article.php?articleid=200, (Johns Hopkins University: 5 May 2003).

——'Tajik Opposition Party Protests Official Repression'. http://www.cacianalyst.org/view_article.php?articleid=996, (Johns Hopkins University: 21 May 2002).

——'Tajik Opposition Party's Registration Rejected'. http://www.cacianalyst.org/view_article.php?articleid=2268, (Johns Hopkins University: 24 July 2005).

——'Tajikistan's Rakhmonov Strengthens Long-term Grip on Power' http://www.cacianalyst.org/view_article.php?articleid=1545, (Johns Hopkins University: 30 June 2004).

——'US Diplomat Raises Human Rights Concerns with Uzbekistan'. http://www.cacianalyst.org/view_article.php?articleid=1786, (Johns Hopkins University: 19 September 2003).

——'Uzbek Human Rights Activist Sentenced'. http://www.cacianalyst.org/view_article.php?articleid=775, (Johns Hopkins University: October 2005).

——'Uzbek President Granted Immunity for Life'. http://www.cacianalyst.org/view_article.php?articleid=1412, (John Hopkins University: 17 June 2004).

——'Uzbekistan Abolishes Media Censorship…'. http://www.cacianalyst.org/view_article.php?articleid=1186, (Johns Hopkins University: October 2005).

——'Uzbekistan Tightens Control Over Groups'. http://www.cacianalyst.org/view_article.php?articleid=2089, (Johns Hopkins University: 12 February 2004).

Chomsky, Noam, *Hegemony or Survival: America's Quest for Global Dominance*, (Henry Holt & Co., 2004).

Chua, Amy, *World on Fire: How Exporting Free-Market Democracy Breeds Ethnic Hatred and Global Instability*, (Bantam Doubleday Dell Publishing Group, 2004).

Cohen, Ariel, Eurasianet, 'Bush Administration Welcomes the Kyrgyz Revolution', http://www.eurasianet.org/departments/insight/articles/eav022905.shtml, (Washington DC: 24 May 2005).

——'After the Uzbekistan Terror Attacks: Don't Sanction, but Press for Democracy', The Heritage Foundation, http://www.heritage.org/Research/RussiaandEurasia/wm496.cfm, (Washington DC: 12 October 2004).

——'Radical Islam and US Interests in Central Asia', Testimony before the Subcommittee on Middle East and Central Asia, Committee on International Relations, US House of Representatives, (Washington DC: 29 October 2003).

Collins, Kathleen, *The Logic of Clan Politics in Central Asia: Its Impact on Regime Transformation.* (Cambridge: Cambridge University Press, 2005).

Cornell, Svante and Niklas Swanstrom, Central Asia-Caucasus Analyst, 'Kyrgyzstan's Revolution: Poppies or Tulips?' http://www.cacianalyst.org/view_article.php?articleid=3323, (Johns Hopkins University: 27 June 2005).

Crawshaw, Steve, 'Getting Away with Murder', *The Guardian.* (London: 31 August 2005).

Daalder, Ivo and James Lindsay, *America Unbound: The Bush Revolution in Foreign Policy*, (John Wiley & Sons, Inc, 2005).

Davis, Matthew, 'Uzbek Crisis Poses Dilemma for US', BBC News, (London: 16 May 2005).

Dawisha, Karen and Bruce Parrott, eds, *Conflict, Cleavage and Change in Central Asia and the Caucasus*, (Cambridge: Cambridge University Press, 1997).

d'Encausse, Helene Carrere, *Islam and the Russian Empire: Reform and Revolution in Central Asia*, (London: I.B. Tauris & Co., 1988).

Department of Defense, 'Secretary of Defense Rumsfeld and Ambassador Purnell Press Conference in Uzbekistan', http://www.defenselink.mil/transcripts/2004/tr20040225-secdef0492.html, (Washington DC: 23 December 2004).

DeWeese, D., *Islamization and Native religion in the Golden Horde*, (University Park, PA: Pennsylvania State University, 1994).

Dubnov, Arkadii, 'Tajikistan', The Rand Corporation, (Washington DC: 1996).

Dudoignon, Stephane and Komatsu Hisao, eds, *Islam in Politics in Russia and Central Asia (Early Eighteenth to Late Twentieth Centuries)*, (London and New York: Kegan Paul, 2001).

Eisenhower Institute's Center for Political and Strategic Studies, 'An Overview of the Ferghana Valley', vol. 1, no. 1, (Washington DC: January 1996).

Eland, Ivan, *The Empire Has No Clothes: US Foreign Policy Exposed*, (The Independent Institute, 2004).

Electionworld. 'Elections in Uzbekistan', http://www.electionworld.org/uzbekistan.htm, (New York: 4 September 2005).

Eurasianet, 'Promoting Media Freedom in Kyrgyzstan Proving More Difficult than Originally Anticipated'. http://www.eurasianet.org/departments/insight/articles/eav110905.shtml, (Washington DC: 6 December 2005).

—— 'Bakiev Wins Landslide in Kyrgyzstan, Courts Controversy with United States'. http://www.eurasianet.org/departments/insight/articles/eav071105.shtml, (Washington DC: 15 August 2005).

——'Kyrgyzstan Local Press Digest', Digest #12, 1–17 July 2000. http://www.eurasianet.org/resource/kyrgyzstan/press_digest/Ky071800.shtml, (Washington DC: July 2000).

——'Kyrgyzstan: Is the President Losing Interest in Constitutional Reform?' http://www.eurasianet.org/departments/insight/articles/eav082505.shtml, (Washington DC: 23 September 2005).

——'Kyrgyzstan's Revolution at Risk'. http://www.eurasianet.org/departments/insight/articles/eav092605.shtml, (Washington DC: 18 October 2005).

——'Presidential Elections in Uzbekistan in 2000'. http://www.eurasianet.org/departments/election/uzbekistan/presanal.htm, (Washington DC: 12 July 2000).

——'Uzbek Government Exerting Pressure on Local NGOs to Close Voluntarily'. http://www.eurasianet.org/departments/civilsociety/articles/eav100405.shtml, (Washington DC: 13 November 2005).

——'Uzbekistan Holds Elections – Without Opposition'. http://www.eurasianet.org/departments/civilsociety/articles/eav12230.htm, (Washington DC: 16 September 2004).

Ferguson, Niall, *Colossus: The Price of America's Empire*, (New York: Penguin Group, 2004).

Fukuyama, Francis, *America at the Crossroads: Democracy, Power and the Neoconservative Legacy*, (New Haven: Yale University Press, 2006).

——*Statebuilding: Governance and World Order in the 21st Century*, (Cornell University Press, 2004).

Gaddis, John Lewis, *Surprise, Security and the American Experience*, (Boston: Harvard University Press, 2005).

Gerges, Fawaz, *The Far Enemy: Why Jihad Went Global*, (Cambridge: Cambridge University Press, 2005).

Gleason, Gregory, *Markets and Politics in Central Asia: Structural Reform and Political Change*, (New York: Routledge, 2003).

Grousset, Rene, *The Empire of the Steppes: A History of Central Asia*, Translated by Naomi Walford, (New Brunswick, NJ: Rutgers University Press, 1970).

Haas, Richard, *The Opportunity*, (New York: Perseus Publishing, 2005).

Haghayeghi, Mehrdad, *Islam and Politics in Central Asia*: (Palgrave Macmillan, 1996).

Hizb ut-Tahrir, 'Demokratiya Sistema Neveriya', http://www.hizb-ut-tahrir.org/russian/htm/democracy.htm, (London: 8 October 2005).

——'Demokratiya Sistema Neveriya', http://www.hizb-ut-tahrir.org/russian/htm/democracy.htm, (London: 12 December 2004).

——'Opori Amerikanskoi Aggressii', http://www.hizb-ut-tahrir.org/russian/htm/esdarat.htm, (London: 30 August 2005).

——'Otveti na Voprosi', http://www.hizb-ut-tahrir.org/russian/htm/democracy.htm, (London: 4 February 2005).

Hopkirk, Peter, *The Great Game: The Struggle for Empire in Central Asia*, (Kodansha Globe, 1994).

Human Rights and Democracy Fund, 'Mission Statement'. http://www.state.gov/g/drl/c7607.htm, (Washington DC: 13 October 2005).

Human Rights Watch, 'Uzbekistan: Andijan Show Trial Ends with Guilty Verdict'. http://hrw.org/english/docs/2005/11/14/uzbeki12020.htm, (New York: 23 November 2005).

——'Burying the Truth: Uzbekistan Rewrites the Story of the Andijan Massacre', vol. 17, no. 6(D). (New York: September 2005).

——'Creating Enemies of the State: Religious Persecution in Uzbekistan', (New York and Washington DC, 2004).

——'Revocation of Accreditation', http://www.hrw.org/reports/1999/Tajikistan/Tajik99n-08.htm#TopOfPage, (New York: 28 November 2004).

——'Tajikistan Report 2002', http://www.hrw.org/wr2k3/europe12.html, (New York: 1 April 2004).

——'Tajikistan Report 1999', http://www.hrw.org, (New York: 21 November 2003).

——'Interview with Maksudjon Husseinov – Tajikistan Report 1999', http://www.hrw.org, (New York: 14 November 2003).

——'Tajikistan Report 1999', http://www.hrw.org, (New York: 16 April 2003).

——'Tajikistan Report 1999', http://www.hrw.org, (New York: 23 March 2002).

——'Dangerous Dealings: Changes to US Military Assistance after September 11', vol. 14, no. 1(G), (New York: February 2002).

——'Tajikistan Report 1999', http://www.hrw.org, (New York: 8 November 2003).

——'US: Military Aid After 9/11 Threatens Human Rights'. http://www.hrw.org/english/docs/2002/02/15/usint3750.htm, (New York: 3 April 2005).

Ikenberry, John G., *America Unrivaled: The Future of the Balance of Power*, (New York: Cornell University Press, 2002).

Institute for War and Peace Reporting, 'Will US Policy Backfire in Central Asia'. http://www.muslimuzbekistan.com/eng/ennews/2004/03/ennews30032004a.html, (21 December 2004).

International Crisis Group, 'Tajikistan's Politics: Confrontation or Consolidation?' (Dushanbe/Brussels: 19 May 2004).

——'Central Asia: Islam and the State', ICG Asia Report no. 59, (Osh/Brussles 10 July 2003).

——'Central Asia: Islamist Mobilization and Regional Security', ICG Asia Report no. 14, 1 (New York: March 2001).

——'Central Asian Perspectives on 11 September and the Afghan Crisis', Osh/Brussels, no. 28, (New York: September 2001).

——Interview with Jantoro Satybaldiev, Mayor of Osh, Kyrgyzstan, (Washington DC: 21 June 2002).

——'Kyrgyzstan's Political Crisis: An Exit Strategy', ICG Asia Report, no. 37, (New York: 19 August 2002).

——'The Failure of Reform in Uzbekistan: Ways forward for the International Community'. ICG Asia Report no. 76, (New York: 11 March 2004).

——'The IMU and the Hizb ut-Tahrir: Implications of the Afghanistan Campaign', Central Asia Briefing, (Osh/Brussels: 30 January 2002).

——'Uzbekistan's Reform Program: Illusion or Reality?' Asia Report no. 46, (Osh/Brussels: 18 February 2003).

International War and Peace Reporting, 'Will US Policy Backfire in Central Asia?' (New York: 30 March 2004).

Jervis, Robert, *American Foreign Policy in a New Era*, (New York: Routledge, 2005).

Kaiser, Bob and Lois Raimondo, 'The Volatile Ferghana Valley', *The Washington Post*, 15 July 2002.

Khadimov, Alisher, Eurasianet, 'Islamic Radical Group Intensifies Underground Activity in Kyrgyzstan', http://www.eurasinet.org/departments/rights/articles/eav041403.shtml, (Washington DC: 18 October 2004).

——'Kyrgyzstan: Organized Opposition and Civil Unrest'. http://www.eurasianet.org/departments/rights/articles/eav121602.shtml, (Washington DC: 30 October 2003).

——'Kyrgyzstan's Unrest Linked to Clan Rivalries'. http://www.eurasianet.org/departments/insight/articles/eav060502.shtm, (Washington DC: 4 January 2004).

Khalid, Adeeb, *The Politics of Muslim Cultural Reform: Jadidism in Central Asia*, (University of California Press, 1999).

Lobe, Jim, Muslim Uzbekistan, 'With Friends Like Uzbekistan'. http://archive.muslimuzbekistan.com/eng/ennews/2004/04/ennews01042004b.htm, (Tashkent: 14 November 2005).

Lorenz, Richard, 'Economic Bases of the Basmachi Movement in the Farghana Valley', In *Muslim Communities Reemerge*, edited by Edward Allworth, (London: 1994).

Lubin, Nancy, 'Who's Watching the Watchdogs?' *Journal of International Affairs* 56, no. 3: 43–58, (2003).

Luong, Pauline Jones, *Institutional Change and Political Continuity in Post-Soviet Central Asia: Power, Perceptions, and Pacts*, (Cambridge: Cambridge University Press, 2002).

Manz, Beatrice, ed., *Central Asia in Historical Perspective*, (Boulder, CO: Westview, 1994).

Matveeva, Anna, 'Democratization, Legitimacy, and Political Change in Central Asia', *International Affairs* 75, no. 1: 23–44, (1999).

McCormick, James, *American Foreign Policy and Process*, (Wadsworth: 2004).

Mead, Walter Russell, *Special Providence: American Foreign Policy and How it Changed the World*, (New York: Taylor & Francis, 2002).

Merry, Robert W., *Sands of Empire: Missonary Zeal, American Foreign Policy and the Hazards of Global Ambition*, (New York: Simon & Schuster Adult Publishing Group, 2005).

Montgomery, David, Euarsianet, 'Religious Persecution and Extremism Fosters Discontent in Central Asia'. http://www.eurasianet.org/departments/rights/articles/pp083002.shtml, (Washington DC: 15 July 2003).

Nazarov, Nazar, Eurasianet, 'Tajikistan Government Resorts to Repression and Intimidation'.

http://www.eurasianet.org/departments/insight/articles/eav042905a.shtml, (Washington DC: 13 October 2005).

O'Kane, John and Touraj Atabaki, eds, *Post-Soviet Central Asia*, (London and New York: Tauris Academic Studies, 1998).

Olcott, Martha Brill, *Central Asia's New States: Independence, Foreign Policy, International Security*: (United States Institute of Peace Press, 1996).

Olimova, Saodat, 'Political Islam and Conflict in Tajikistan', CA & CC Press, (Sweden, 2005).

O'Loughlin, John and Altinay Kuchukeeva, 'Civic Engagement and Democratic Consolidation in Kyrgyzstan', *Post-Soviet Geography and Economics* 44, no. 8: 557–587 (2003).

Peterson, Scott, 'Amid Reforms, Muslims Still Under Fire in Ferghana Valley', Christian Science Monitor, (Washington DC: 19 July 2002).

Pillar, Paul, *Terrorism and US Foreign Policy*, (Washington DC: Brookings Institution Press, 2003).

Polat, Abdumannov, 'The Islamic Revival in Uzbekistan: A Threat to Stability?' In *Islam in Central Asia: An Enduring Legacy or an Evolving Threat?*, edited by Roald Sagdeev and Susan Eisenhower, 39–57, (Washington: Center for Political and Strategic Studies, 2000).

Radio Free Europe/Radio Liberty, 'Tajikistan Votes 2005', http://www.rferl.org, (Washington DC: 3 December 2005).

—'Profile: Feliks Kulov', http://www.rferl.org/featuresarticle/2005/03/ec9e9641-1632-4981-a435-21505d93a517.html, (Washington DC: 16 November 2003).

—'Tajik Elections', http://www.rferl.org/specials/Tajikelections/timeline2000.asp, (Washington DC: 12 January 2004).

—'Tajik Elections', http://www.rferl.org/specials/Tajikelections/timeline2001.asp, (Washington DC: 03 January 2005).

—'Tajik Elections', http://www.rferl.org/specials/Tajikelections/timeline2004.asp, (Washington DC: 09 February 2005).

—'Tajikistan Votes 2005', http://www.rferl.org, (Washington DC: 17 November 2005).

—'Tajikistan Votes 2005', http://www.rferl.org, (Washington DC: 25 November 2005).

Rotar, Igor, 'Enlightened Islam, Uzbek-style: Islam Karimov is getting rid of his most dangerous rivals', *Prism* 4, no. 13, (Washington DC: 26 June 1998).

Roy, Olivier, 'The Foreign Policy of the Central Asian Islamic Renaissance Party', Council on Foreign Relations, (New York, 2000).

Ryn, Claes, *America the Virtuous: The Crisis of Democracy and the Quest for Empire*, (Transaction Publishers, 2003).

Saidazimova, Gulnoza, Eaurasianet, 'Kyrgyzstan: New Leadership Drawn from the Old Guard', http://www.eurasianet.org/departments/insight/articles/pp032505a.shtml, (Washington DC: 15 November 2005).

Saroyan, Mark, *Minorities, Mullahs, and Modernity: Reshaping Community in the Former Soviet Union*. Edited by Edward W. Walker, (Berkeley, CA: International and Area Studies, UC Berkeley, 1997).

Sestanovich, Stephen, 'Promoting Democracy', *Journal of International Affairs* 56, no. 3: 149–156 (2003).

Sharansky, Natan and Ron Dermer, *The Case for Democracy: The Power of Freedom to Overcome Tyranny and Terror*, (Public Affairs, 2004).

Shirazi, Habibollah Abolhassan, 'Political Forces and their Structures in Tajikistan', *Central Asian Survey* 16, no. 4 (1997): 611–622.

Shukurov, Sharif and Rustam Shukurov, *Tsentral'naia Aziia: Opyt istorii dukha*, (Moskva: Tsentr strategicheskogo planirovaniia Orenburgskoi oblasti, 2001).

Umarov, Noor, Central Asia-Caucasus Analyst, 'Does Tajikistan's Parliamentary Elections Spell the End of the UTO?' http://www.cacianalyst.org/view_article.php?articleid=179, (Johns Hopkins University: 2004).

United Nations Development Program, 'Environment and Security: Transforming Risks into Cooperation', http://www.envsec.org/centasia/maps.php, (New York: 13 November 2005).

US House of Representatives, Committee on International Relations, Subcommittee on the Middle East and South Asia, 'US Policy on Central Asia'. http://commdocs.house.gov/committees/intlrel/hfa72975.000/hfa72975_0f.htm, (Washington DC: 12 June 2003).

The United States Mission to the European Union, 'Assistant Secretary of State Jones on US Policy in Central Asia'. http://www.useu.be/Terrorism/USResponse/Dec1401JonesUSPolicyCentralAsia.html, (Washington DC: 12 January 2004).

Voll, John O., 'Central Asia as Part of the Modern Islamic World', In *Central Asia in Historical Perspective*, edited by Beatrice Manz, 62–81. (Boulder, CO: Westview Press, 1994).

Zakaria, Fareed, *The Future of Freedom: Illiberal Democracy at Home and Abroad*, (New York: W.W. Norton & Co., 2003).

Zunes, Stephen and Richard Falk, *Tinderbox: US Foreign Policy and the Roots of Terrorism*, (Common Courage Press, 2002).

Index